GOANNA

Australian Natural History Series

Series Editor: Professor Terry Dawson

The function of this series of titles is to make accessible accurate scientific information, complemented by high quality illustrations, on a wide variety of Australian animals. Written and illustrated by trained researchers and scientists, they are intended for students and biologists at both secondary and tertiary levels and, in general, for readers with a serious interest in animals and the environment.

Interested authors should contact the New South Wales University Press or Professor Dawson, School of Zoology, University of New South Wales, PO Box 1, Kensington 2033.

Books planned in the series include the kangaroo, the echidna and the dingo.

Books published include:
The Platypus, Tom Grant Illustrated by Dominic Fanning
Little Penguin, Colin Stahel & Rosemary Gales Illustrated by Jane Burrell
The Koala, Anthony Lee & Roger Martin Illustrated by Simon Ward
The Wombat, Barbara Triggs Illustrated by Ross Goldingay
The Lyrebird, Pauline Reilly Illustrated by Peter Chambers
Goanna, Dennis King & Brian Green Illustrated by Frank Knight, Keith Newgrain and Jo Eberhard

GOANNA

THE BIOLOGY OF VARANID LIZARDS

Dennis King & Brian Green

Illustrated by
Frank Knight, Keith Newgrain
and Jo Eberhard

For Ruth and Iris

Published by
NEW SOUTH WALES UNIVERSITY PRESS
PO Box 1 Kensington NSW Australia 2033
Telephone (02) 398 8177
Fax (02) 398 3408

First published in 1993

National Library of Australia
Cataloguing-in-Publication entry:
King, Dennis, 1942–
Goanna. The biology of varanid lizard

Bibliography.
Includes index.
ISBN 0 86840 093 9.

1. Monitor lizards. I. Green, Brian, 1942– . II. Title.

597.95

Available in North America through:
International Specialised Book Services
5804 N.E. Hassalo Street
Portland Oregon 97213-3644
United States of America

Typeset by Character & Caps, North Sydney NSW
Printed by Southwood Press, Marrickville NSW

Photograph on previous page: *Varanus rosenbergi*
Photo by R.E. Johnstone

CONTENTS

PREFACE

Why have we written a book about goannas? Apart from the obvious attraction of the unbounded wealth soon to befall us in royalties, we have long felt that a general text on the biology of goannas, for naturalists and students alike, was needed. Goannas, otherwise known as varanids or monitors, are a typically Australian group of lizards. Of the 40 or so described varanid species in the world, 25 of them (and several more as yet undescribed species) are found in Australia. Most of the published information is contained in scientific journals, and many of these papers are written in a range of languages other than English. As such, this information is not readily available to, or easily understood by, the general public in Australia. We have therefore attempted to draw together and condense the more interesting aspects of varanid biology, and have tried to present this information in an easily readable form.

Goannas are different from most other lizards with respect to their ecology, behaviour, anatomy and physiology. Many of these differences have enabled them to flourish in harsh environments and to serve in the role of top predators in many regions, especially in Australia. Many goannas are quite large; the Komodo dragon is the largest lizard in the world and has been reported to kill and devour humans. Their attributes have fostered great interest in the comparative anatomy and physiology of goannas, to the point where we know more about these subjects than about the general ecology of these lizards. Hopefully, this book will stimulate much needed research into their ecology and behaviour, to help redress this imbalance.

Much of the information contained in this book is the work of other researchers. The list of suggested further reading at the end of this book indicates the sources of much of this information. We are particularly grateful for the useful conversations and correspondence we have had with Hans Otto Becker, Wolfgang Böhme, Mike Calver, Tristram Davies, Gil Dryden, Maren Gaulke, Hans-Georg Horn, Gary Packard, Harvey Pough, Tom Tozer and Phil Withers.

The line drawings were prepared by Jo Eberhard, apart from Figures 1, 2, 7, 16, 17, 22, 23, 24, 25 and Table 16 which were prepared by Frank Knight. The graphs and figure annotations were prepared by Keith Newgrain.

Various drafts of the book were reviewed and criticised by Ruth Allen, Terry Dawson, Jo Eberhard, Allen Greer, Rick Shine and Jill Wayment. They provided many helpful suggestions on ways to clarify and improve the book. However, the responsibility for any factual errors or gross grammatical faults that still remain must be accepted by Max King.

PROLOGUE

She moved into the small clearing and stopped in the spring sunlight. Head raised, tongue flickering, alert, she patiently searched the surrounding bush and waited. The only sounds were those from a gentle breeze rustling the gum leaves, and the distant chatter of parrots. Quietly, carefully, she started across the clearing, moving in a slow and undulating fashion. She examined and sniffed a small burrow, then turned and left it. On the far side of the clearing lay a rotting log – her next point of investigation. Again her tongue flickered, sampling the air for the scent of food. Strong claws tore at the log, displacing a small skink from its shelter, but after a brief chase and a flash of sharp teeth it was seized. The skink struggled, but her teeth had pierced its body and the violent shaking of her head had ensured its sudden death. The skink's body was turned so that the head pointed towards the back of her jaws, and with a few twists of her neck it disappeared down her throat. All was still in the surrounding scrub. Her long, scaly body moved silently away and the hunt resumed. Spring had come to the island and with its arrival, the goanna had begun to hunt again.

CHAPTER 1

INTRODUCTION

What is a goanna? It is the common name in Australia for a lizard in the family Varanidae. However, in some parts of the country it is a name which is also given to any species of large lizard and is often incorrectly used for members of another family, the Scincidae. 'Goanna' is believed to be a corruption of the name 'iguana', which belongs to a different family of lizards found mainly in South America.

Goannas are distributed over a large part of the southern hemisphere, from Africa through southern Asia to Australia. In other countries they are commonly known as 'varanids' or 'monitor lizards'. The name 'goanna' is firmly entrenched in the terminology of Australians. No matter what scientists or others may call them, or how many kinds of lizards this common name might be applied to, it is a name which will continue to be used, even if only in Australia. The names 'goanna', 'monitor' and 'varanid' are used interchangeably in this book.

All varanids can be easily recognised as belonging to the same family of lizards. They are all insectivores or carnivores and all have a similar appearance, although they differ in size and body proportions. The body form of a typical monitor is shown in Figure 1.

Varanids have evolved into a number of species and have become specialised for some very different lifestyles. Some species have become adapted to an amphibious mode of life, some are arboreal and others are adapted to living in some of the harshest deserts of the world. Others occur in mangrove swamps or dense rainforests, while a close relative, the earless monitor (*Lanthanotus borneensis*), has adopted a subterranean lifestyle in the jungles of Borneo.

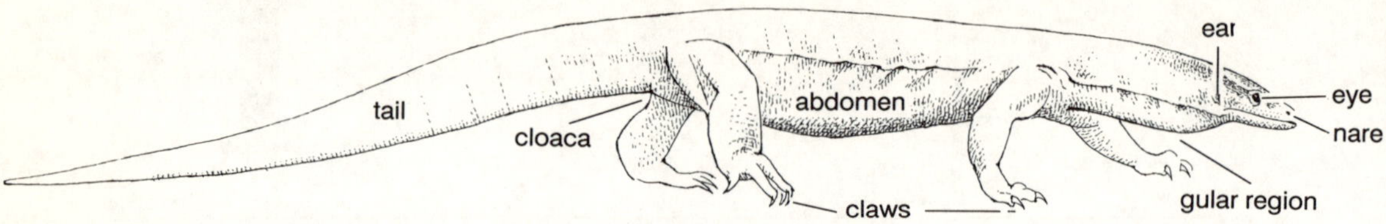

Figure 1
General morphology of a varanid

Varanids range in length from the pygmy species in Australia, with adult total lengths of as little as 0.21 metres in *Varanus brevicauda*, to 3 metres for the Komodo dragon (*Varanus komodoensis*) of eastern Indonesia. An arboreal species in New Guinea, *Varanus salvadorii*, has been reported to reach a length of 4 metres, but this has not been verified. In weight they range from as low as 17 grams for adult *Varanus brevicauda* to at least 70 kilograms, and possibly over 100 kilograms, for the Komodo dragon. An unconfirmed weight of 140 kilograms has been reported for a Komodo dragon which was shot in 1912, soon after the species first became known. The Komodo dragon is the largest lizard in the world today (Plate 2), but even it would be dwarfed by the extinct species *Megalania prisca* (Figure 2) which was found over much of inland eastern Australia up to about 30,000 years ago. It reached a total length of 7 metres and is estimated to have weighed 650 kilograms. Early Aboriginal hunters may have encountered these awesome creatures while hunting in the Australian bush. The largest species found in Australia today (*Varanus giganteus*) is only about 2 metres long.

The closest living relatives of the Varanidae are believed to be the Lanthanotidae, or earless monitors, from Borneo, and the Helodermatidae, or beaded lizards, of Mexico and Central America. The lanthanotids are represented by a single species, the earless monitor (*Lanthanotis borneensis*) which has adopted a burrowing and apparently wholly subterranean existence. There are two species of helodermatids, the gila monster *Heloderma suspectum* and the beaded lizard *Heloderma horridum*. They are slow-moving animals which feed on invertebrates and birds' eggs and are the only species of lizards which are known to be venomous. These two groups of lizards and the varanids are all thought to be related to the extinct mosasaurs, an aquatic group of large reptiles which probably preyed upon fish.

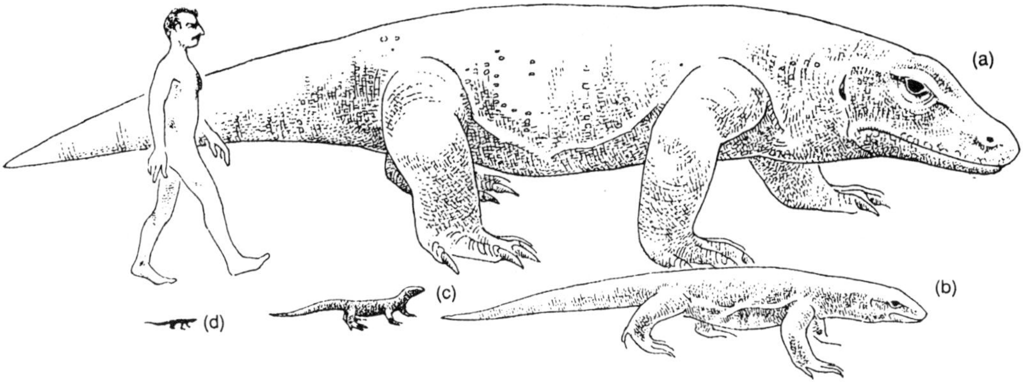

Figure 2
Relative sizes of some varanids: a. *Megalania prisca*; b. *V. komodoensis*; c. *V. rosenbergi*; d. *V. brevicauda*

The genus *Varanus* is most widespread in Australia, where about 30 of the world's approximately 40 species are found. There are 14 species in Asia and only 4 species in the whole of Africa. The world distribution of varanids extends as far east as the Philippines, as far west as the west coast of Africa, and from the Caucasus in the north to the southern regions of Australia, with the exception of Tasmania (Figure 3).

The diversity and abundance of varanid lizards in Australia may be due to the absence of competition from mammalian carnivores during their evolutionary history, as compared to the abundant and diverse range of carnivorous mammals found in Asia and Africa. As many as 10 different species of goannas may occur in the same general area in parts of Australia, but competition between them is minimised in several ways.

The Australian goanna that has been studied in most detail is Rosenberg's monitor, *Varanus rosenbergi* (Plate 1). It can reach a total length of 1 metre and weigh up to 2 kilograms. It occurs along much of the south coast of Australia (Figure 4). Specimens which appear to belong to this species have also been collected near Canberra and Sydney but their identity has not yet been confirmed. This book deals mainly with the biology of this species and compares it with other species of varanids where appropriate.

Rosenberg's monitor is very abundant on Kangaroo Island,

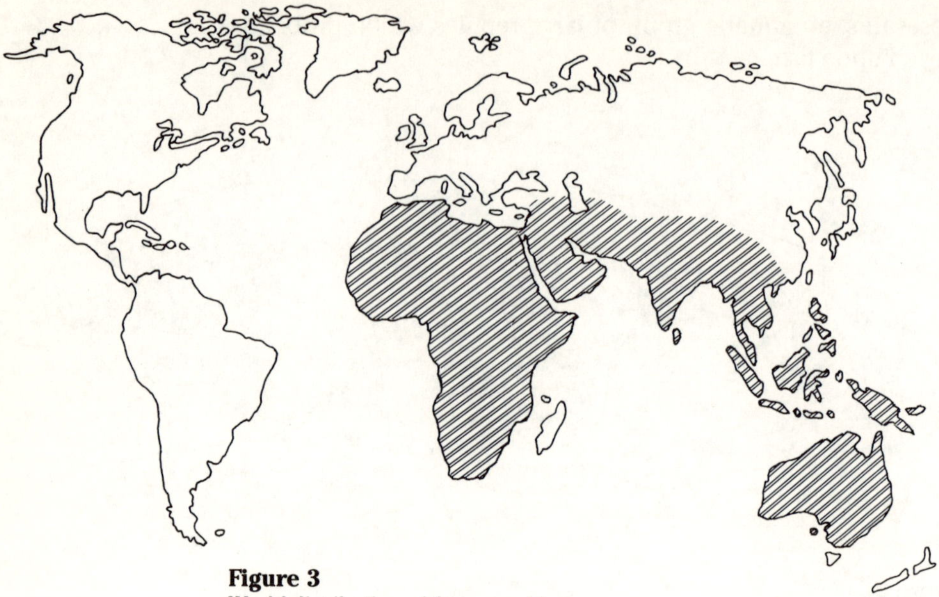

Figure 3
World distribution of the genus *Varanus*

which lies off the coast of South Australia and has been separated from the mainland for 6000 to 8000 years. *Varanus rosenbergi* is a medium-sized goanna that is the largest terrestrial predator on Kangaroo Island. There are fossilised remains of larger marsupial carnivores at several sites on the island, but these species no longer occur there. Adult *Varanus rosenbergi* are now killed only by humans, vehicles and raptorial birds such as wedge-tailed eagles.

Prior to 1957, *Varanus rosenbergi* was referred to as 'a melanistic form' of *Varanus gouldii*. In that year it was described as a distinct subspecies of Gould's monitor, *Varanus gouldii*. In a recent taxonomic revision of the varanid lizards found in Western Australia, however, it was recognised as being sufficiently distinct from *Varanus gouldii* to warrant the status of a full species, and subsequent taxonomic studies have provided further evidence which supports that decision.

Kangaroo Island, at 35° 45' S, 138° 15' E, is almost as far from the equator as any location where varanid lizards can be found (Figure 3). The lace monitor, *Varanus varius*, occurs slightly further south in some areas of the state of Victoria, Australia. The desert monitor, *Varanus griseus*, is the northernmost species of varanid. It occurs between 35° and 40° north of the equator in areas of the southernmost republics of the former USSR.

Figure 4
Distribution of *V. rosenbergi*

CHAPTER 2

TAXONOMY AND PHYLOGENY: Who's Who, Their Relationships and Origins

Varanid-like reptiles first appear in the fossil record during the Cretaceous period, about 100 million years ago. After this period a wide range of both terrestrial and aquatic forms appeared. However, most of them became extinct by the end of the Palaeocene era (about 60 million years ago). These fossil varanoids are found throughout Europe, Asia and North America, but the fauna of the Cretaceous period is not known well enough for the evolutionary history of the varanids to be accurately described.

The earliest fossil form that can definitely be ascribed to the Varanidae is *Telmasaurus*, found in Upper Cretaceous deposits in Mongolia. Another varanoid family, the Necrosauridae, was present in Europe from the Upper Palaeocene up to the Oligocene, about 65 to 26 million years before the present (BP). The genus *Varanus* first appeared in the fossil deposits of the Miocene (15 million years BP) in eastern Europe, Africa and Australia, and it is likely that the genus originated in Asia and radiated outwards into Africa and Australia.

The first thorough attempt to classify the living species, currently all placed in the genus *Varanus*, was that of Robert Mertens, who worked in Germany. His work, published in 1942, was severely hampered by the Second World War, since he did not have access to specimens in museums in many other countries. Mertens placed the 24 species of his original classification into eight subgenera, and some of the species were divided into several subspecies. In two subsequent studies he proposed two further subgenera, to give a total of ten. Mertens' classification was based almost entirely on the osteology (study of the bones) and the external morphology of the animals. It has since been modified extensively by workers who have used a range of techniques and who have also had access to larger collections of specimens from wider geographic areas. Most of the subgenera named by Mertens are now being revised and altered.

One technique used to investigate taxonomy and phylogeny is to examine and compare the number and morphology (shape and size) of the chromosomes. While the number of chromosomes in all varanids is constant with a diploid number of 40, their arrangement does differ and the species can be placed in six groups on the basis of these characteristics (Figure 5). These groups are not consistent with the subgenera of Mertens, as one of his groups contains members of four sub-genera, and members of another subgenus belong to three different chromosome groups. This suggests the need for a new subgeneric division as proposed in Table 1.

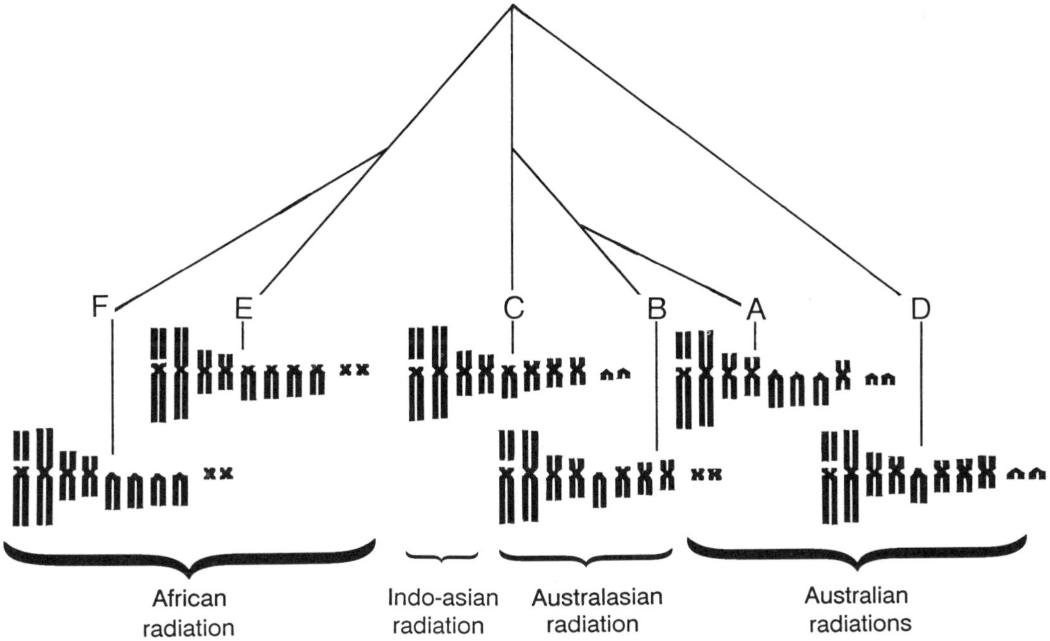

Figure 5
Chromosomal shape and relationships of varanids

Another way of studying the relationships of different species is to compare the rates at which proteins extracted from body tissues move in a medium in response to an electric charge (electrophoresis). Such electrophoretic studies on blood, muscle and liver from varanids show relationships between the species similar to those found by chromosomal studies.

Table 1
Proposed new subgeneric groupings of *Varanus* based on micro-complement fixation data.

Species	Subgenus
albigularis	*Polydaedalus*
exanthematicus	
niloticus	
yemenensis	
griseus	*Psammosaurus*
bengalensis	*Empagusia*
dumerili	
flavescens	
indicus	*Euprepiosaurus*
karlschmidti	
prasinus	
rudicollis	*unnamed*
salvator	
salvadorii	*Papusaurus*
olivaceus	*Philippinosaurus*
acanthurus	*Odatria*
baritji	
brevicauda	
caudolineatus	
eremius	
gilleni	
glauerti	
glebopalma	
kingorum	
mitchelli	
pilbarensis	
primordius	
scalaris	
semiremex	
storri	
timorensis	
tristis	
giganteus	*Varanus*
gouldii	
komodoensis	
mertensi	
panoptes	
rosenbergi	
spenceri	
varius	

Another study used an immunological method called micro-complement fixation (MCF) to compare the structure of the protein albumin in different species. This method assumes that certain proteins mutate at a fixed rate and can therefore be used as a molecular clock to estimate the degree of relatedness between species and when species diverged from each other.

MCF analysis of varanids has produced similar results to the chromosomal studies.

These recent studies indicate that there are at least eight recognisable subgenera, although one technique suggests that an extra subgenus (*Philippinosaurus*) may exist, as was proposed by Mertens. Four of these subgenera are entirely different from those proposed by Mertens. The remainder are essentially the same, with deletions or additions of one or two species to two of them. The three unaltered subgenera (*Philippinosaurus, Papusaurus* and *Psammosaurus*) each consist of a single species. All those consisting of two or more species have been altered in some way to produce new groups of species (Table 1).

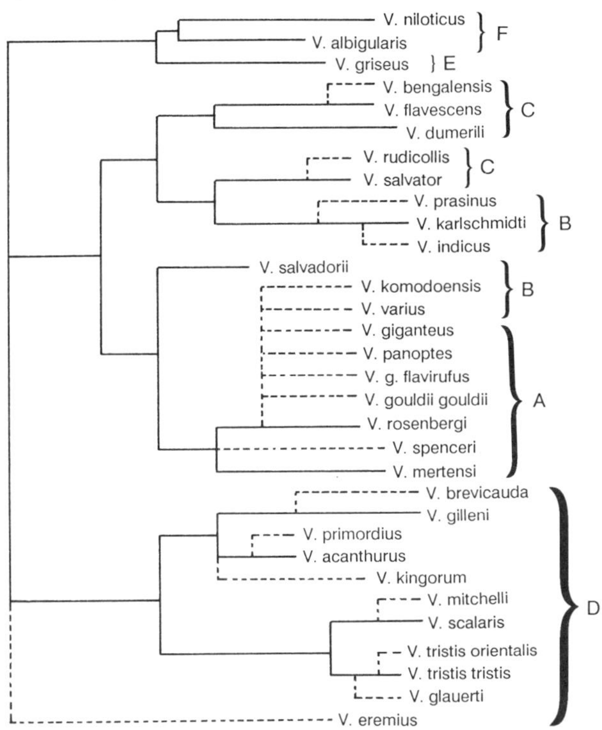

Figure 6
Proposed varanid phylogeny based on MCF data

The phylogeny suggested by the MCF study, which agrees very closely with the results of the chromosome morphology study is shown in Figure 6. Such a phylogeny indicates that the family originated in southeast Asia and radiated east and southwards into Australia and westwards into the Indian subcontinent, the Middle East and Africa. There were several radiations into Australia during the past 40 million years, following the break-up of the Gondwana supercontinent. Between then and the Miocene, Australia moved northwards and eventually came into close contact with southeast Asia. Some species of varanids have moved from Australia to New Guinea, or vice versa, in more recent times. Figure 7 shows the distributions of the proposed new subgenera.

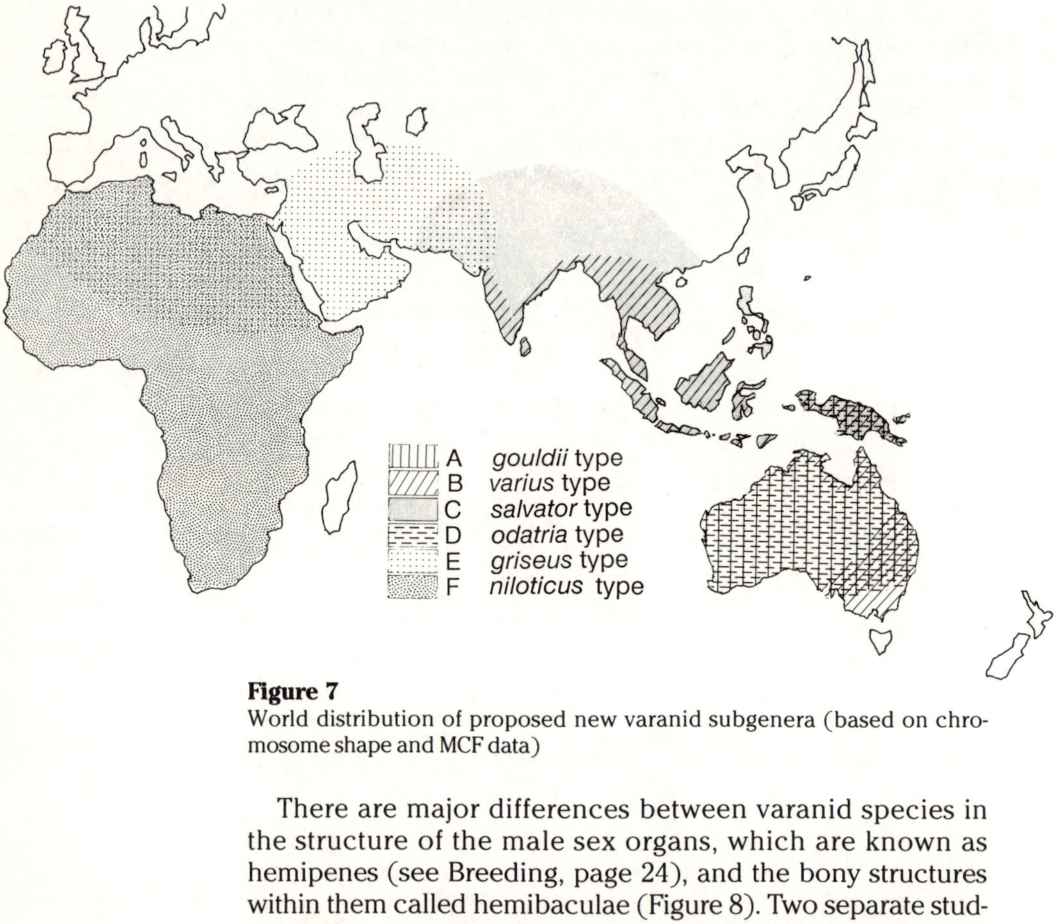

Figure 7
World distribution of proposed new varanid subgenera (based on chromosome shape and MCF data)

There are major differences between varanid species in the structure of the male sex organs, which are known as hemipenes (see Breeding, page 24), and the bony structures within them called hemibaculae (Figure 8). Two separate studies of these organs, and another comparing lung morphology, provide further support for revising the taxonomic groupings proposed by Mertens.

While there is close agreement between recent studies in the species composition of the subgenera, they differ in their interpretation of the ways in which the subgenera relate to each other. For example, the inferred relationships between the Australian subgenus *Varanus* and three Asian subgenera (*Empagusia, Euprepriosaurus* and one unnamed) do not agree. Further studies using other techniques will be necessary to resolve this issue.

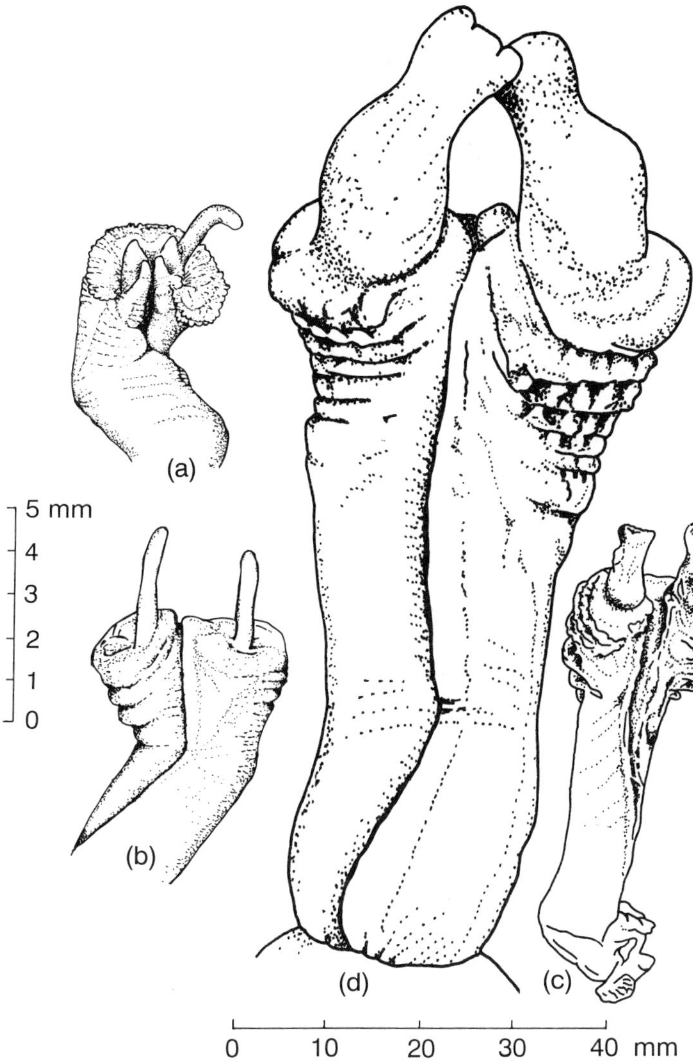

Figure 8
Hemipenes of some varanids; a. *V. gilleni*; b. *V. primordius*; c. *V. gouldii*; d. *V. komodoensis* (after Böhme 1988)

The currently recognised species of *Varanus* are shown in Table 2. Further taxonomic studies are certain to alter this list. The names of some existing species will be changed and some newly recognised. Species will be described and named within the next few years (Plate 3). The present situation, where all species are placed in the single genus *Varanus*, will probably also change. Biochemical studies of some other reptiles have indicated the need to reclassify a number of groups and this has been carried out following re-examination of morphological characteristics. Results of the chromosomal, biochemical and morphological studies mentioned above suggest that the varanids consist of several different genera, so the elevation of the proposed subgenera to full generic status should be seriously considered and investigated.

Table 2
Currently recognised species of *Varanus* and probable changes in taxonomic status

Species	Distribution	To be changed	Probable change
albigularis	Africa		yes
exanthematicus	Africa		yes
niloticus	Africa		
yemenensis	Arabian peninsula		
griseus	North Africa and Middle East		
bengalensis	Asia		
dumerili	Asia		
flavescens	Asia		
komodoensis	Asia (Indonesia)		
olivaceus	Asia (Philippines)		
rudicollis	Asia		
salvator	Asia		yes
timorensis	Asia (Indonesia)		
indicus	Asia and New Guinea	yes	
karlschmidti	New Guinea		yes
prasinus	New Guinea and Australia		yes
salvadorii	New Guinea		
acanthurus	Australia		yes
baritji	Australia		
brevicauda	Australia		
caudolineatus	Australia		
eremius	Australia		
giganteus	Australia		
gilleni	Australia		
glauerti	Australia		
glebopalma	Australia		
gouldii	Australia	yes	
kingorum	Australia		
mertensi	Australia		
mitchelli	Australia		
panoptes	Australia and New Guinea	yes	
pilbarensis	Australia		
primordius	Australia		
rosenbergi	Australia		
scalaris	Australia and New Guinea		yes
semiremex	Australia		
spenceri	Australia		
storri	Australia		
tristis	Australia		yes
varius	Australia		yes

CHAPTER 3

FEEDING

DIET

The diet of lizards is studied either by identifying traces of prey left in faeces or by examining stomach contents. As reptiles generally do not chew their food well, if at all, it can be readily identified. Stomach contents may be examined by dissecting dead, preserved museum specimens or by flushing the stomachs of live lizards, to remove the food they have eaten. By either or both of these methods, the stomach contents of all Australian varanid species have been examined.

All varanids which have been studied are wholly or largely carnivorous, although an arboreal species from the southern Philippines (*Varanus olivaceus*) is partly herbivorous. The invertebrate portion of the diet, which is mainly snails, varies from month to month but can reach up to 95 percent of the total. *Varanus olivaceus* also feeds on a variety of fruits, particularly between May and September, and is the only goanna known to do so. Adult *Varanus olivaceus* eat approximately equal amounts of plant and animal foods over the course of a year, but juveniles eat much more animal than plant food.

Australian pygmy monitors such as *Varanus acanthurus, Varanus brevicauda* and *Varanus glauerti* prey almost exclusively on invertebrates. The larger species, such as *Varanus giganteus, Varanus rosenbergi* and *Varanus varius* also include large items in their diet, such as mammals, birds and reptiles, including their eggs. Very large prey items are not swallowed whole but are eaten in pieces sliced from the prey carcasses. Only the larger species are powerful enough to utilise this source of food. They also feed on carrion from large prey.

When hunting for food, goannas walk with a characteristic swinging gait with their snout held close to the ground. The long forked tongue flickers in and out, transferring odours to sensory organs (Jacobson's organs). By this means, they can rapidly locate hidden prey, even if it is underground. Goannas use their strong claws as well as their snout for digging out prey.

The diet of *Varanus rosenbergi* on Kangaroo Island is shown in Figure 9a. It contains a large number of invertebrates, but vertebrates form 65 percent of the volume of their food. The majority of these vertebrates are probably scavenged from animals killed on the roads. Large species of goannas appear to eat whatever prey or carrion they encounter, as long as it is of suitable size. *Varanus rosenbergi* has a wide-ranging foraging pattern which includes many habitats, ranging from dense bush to open beaches. It often uses taste and smell when searching for hidden prey, and eats many small items of prey. This pattern of foraging probably resembles that of primitive varanids.

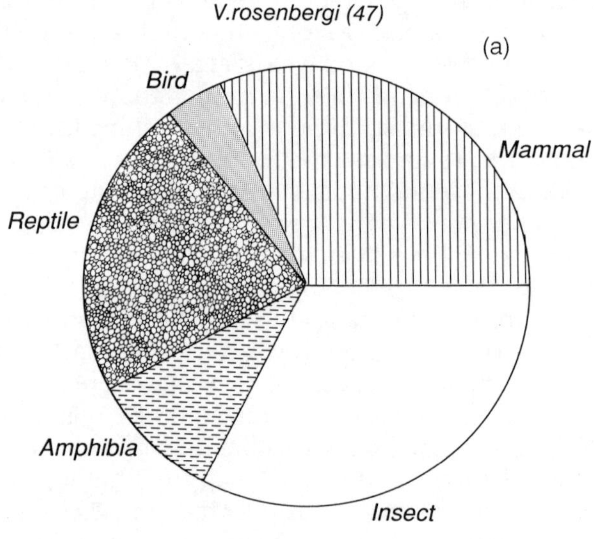

Figure 9(a)
Diet of *V. rosenbergi* (% volume)

The diet of another large varanid from southern Australia (*Varanus varius*), which uses similar habitats, has also been studied (Figure 9b). Arthropods are represented to a similar extent as in the diet of *Varanus rosenbergi*, but mammals are more common in its diet and reptiles are eaten less often. *Varanus varius* eats food which is generally larger than that eaten by *Varanus rosenbergi*, but even the largest animals eat small prey when they encounter it.

Some smaller varanids, such as *Varanus gilleni*, have diets as varied as those of *Varanus rosenbergi* and *Varanus varius* in the content of small prey species (Figure 9b), but they do not eat large vertebrates. However, the larger species (such as *Varanus komodoensis* and *Varanus giganteus*) which seem to rely more on a sit-and-wait hunting strategy than on a wide-foraging one, have few invertebrates in their diets. Presumably invertebrates provide insufficient energy to warrant the emergence of these lizards from their place of ambush. Large Komodo dragons hunt on the ground, and often wait beside game trails. They immobilise passing prey, including Asian buffalo, by seizing them by the leg and tearing their hamstring tendons. They can then kill the crippled prey by tearing at the throat and belly, as they do with their usual smaller prey, such as pigs and deer. Carrion also forms a substantial part of their diet. Small Komodo

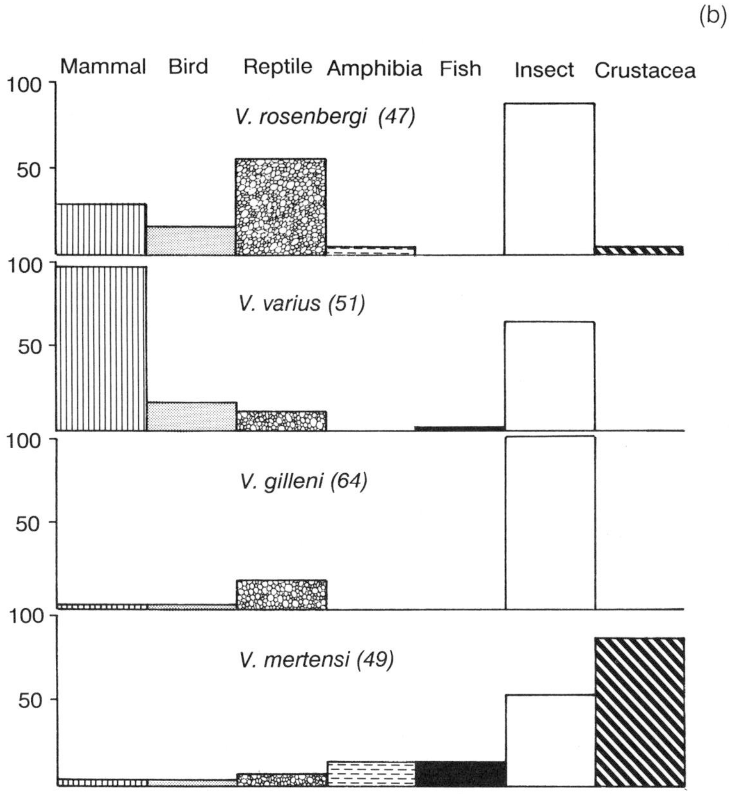

Figure 9(b)
Diet of four varanid species (% occurrence)

dragons usually hunt in the trees and feed mainly on small lizards (skinks and geckoes) and invertebrates, while medium sized animals feed predominantly on birds and rodents. Ambushing is not a typical hunting strategy for varanids, but it does occur in *Varanus komodoensis* and may also occur in some other species, such as *Varanus acanthurus*.

Eggs of other reptiles (turtles, crocodiles, lizards) or of birds are readily eaten by several species of goannas, including *Varanus giganteus, Varanus tristis* and *Varanus varius* in Australia, *Varanus indicus* on Guam, *Varanus salvator* in Asia and *Varanus niloticus* in Africa. Eggs may sometimes form a major portion of the diets of these species. Cannibalism occurs in several varanid species (*Varanus rosenbergi, Varanus gouldii, Varanus giganteus, Varanus komodoensis* and *Varanus bengalensis*). Occasionally it involves feeding on carrion, but predation by wild goannas on live members of their own species has been observed in, for example, *Varanus komodoensis* and *Varanus giganteus*.

The scope of the diet is influenced by season, the location of the animal and by prey diversity and abundance. Varanids opportunistically use locally abundant types of prey. As most of the energy used by varanids in hunting is spent locating prey, rather than in subduing it, a wide range of prey items is acceptable to goannas. Should a small item such as a beetle or a cockroach emerge from a burrow or from under a log, it will be eaten as readily as larger prey such as a lizard or small mammal.

Goannas forage widely and their attention is often engaged in activities such as digging for prey. They thereby expose themselves to predation more than would an ambush predator. Smaller goannas are particularly vulnerable.

Some varanids, including the Australian species *Varanus mertensi* are highly amphibious, mainly capturing and feeding on fish, crustaceans or molluscs in the water (Figure 9b). *Varanus mertensi* (Plate 5) swim well and they have been seen walking underwater, along river bottoms, searching for prey. However, not all the species commonly found near water eat aquatic prey. The diet of *Varanus indicus* is predominantly lizards and the most important food of *Varanus salvator* is grasshoppers and crickets. Other varanids (for example, *Varanus tristis* and *Varanus varius*) are partly or completely arboreal and include eggs or nestlings of tree-nesting birds in their diet.

However, most species forage on the ground and pay particular attention to potential hiding places such as burrows or surface debris under which prey may shelter. Varanids such as *Varanus eremius, Varanus komodoensis* and *Varanus bengalensis* apparently learn to recognise good locations for finding food and regularly return to them during their foraging. These locations include resting and nesting areas of many types of prey. This can lead to dietary specialisations among some monitors, as can the high abundance of certain prey types in their particular habitats.

Some aspects of the anatomy, physiology and behaviour of varanids differ from those of other lizards and allow them to forage differently. These include the structure of some organs, high body temperatures and efficient respiratory systems which enable them to maintain high activity levels. These traits enable the largest species to pursue, capture and eat large vertebrates, but most small or medium-sized species generally feed on invertebrates. The varanids are probably descended from medium sized lizards which fed predominantly on insects which they encountered as they foraged widely throughout their surroundings. Present day varanids may also encounter vertebrates less frequently than invertebrates as they forage, or the smaller species may not be powerful enough to subdue most vertebrates.

FOOD DETECTION

The snout and nasal sacs

The snout of a varanid is important in hunting for prey. Each external nostril (nare) opens into a part of the skull called the nasal capsule, which consists of a number of chambers (Figure 10). This capsule contains the Jacobson's organs.

The nasal capsule in varanids is narrow, elongated and cartilaginous. In the anterior region it is covered only by skin and connective tissue, whereas in other reptiles it is covered by bone. The roof of the posterior region is continuous cartilage in varanids.

The nasal tubes lead from the nares to the anterior chambers, which are not sensory organs, being lined with flattened, non-sensory cells. Spongy tissue surrounds the nasal tubes and the anterior chambers and provides a mechanism for closing off the nasal passages to protect the olfactory chambers. The olfactory chamber is less extensive in varanids than in most other lizards. It is lined with sensory olfactory cells. The olfactory chambers connect with the mouth via internal nares (choanae).

The position of the nares on the snout varies between species. This may be of adaptive significance for goannas which occupy different types of environments. It has been suggested that those species that have nares close to the eyes may forage through litter which they disturb with their snouts.

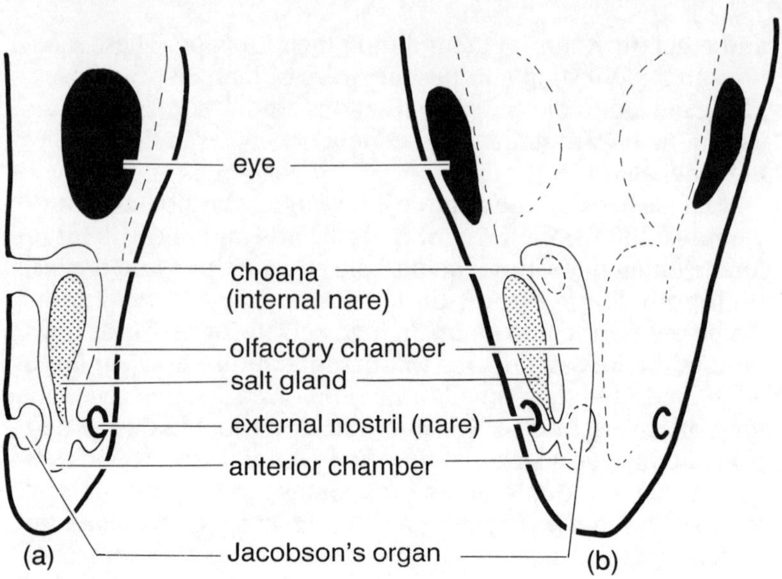

Figure 10
Morphology of the snout region; a. side view; b. top view

Jacobson's organs

In addition to the olfactory chambers, lizards and snakes possess a pair of special sense organs called Jacobson's organs, which are also located in the nasal capsules above the roof of the mouth (Figure 10). Their large size in the Varanidae suggests that they play an important part in the sensory capabilities of goannas. These organs are connected to the roof of the mouth and the nasal sacs by ducts. Jacobson's organs enable a lizard or snake to detect scent particles which have been collected and transferred to the roof of the mouth by the tongue. Jacobson's organs contain sensory cells similar to those found in the olfactory chambers. These sensory cells are connected to the accessory olfactory bulb of the brain by the vomeronasal nerves. Jacobson's organs supplement the olfactory chambers as organs of smell, and are also important in social interactions.

FOOD HANDLING

The tongue and hyoid apparatus

In its primitive form the main role of the tongue in reptiles was the manipulation of food. In most living species of lizards and snakes the tongue is used to manipulate food in the mouth during feeding and also to convey scent particles to the Jacobson's organs. These species are able to protrude the tongue further from the mouth than more primitive lizards and they have more complex Jacobson's organs. In varanids, however, the tongue is modified to fill a sensory role for use in locating prey and in social behaviour. The Varanidae are the only family of lizards which uses the tongue exclusively for sensory functions. Snakes' tongues are similar in appearance and are also used only for sensory purposes. This has led to suggestions of a possible close evolutionary relationship between varanids and snakes. However, the musculature of their tongues differs substantially, and studies of a number of other characteristics indicate that, although snakes may be more closely related to varanids than to other lizards, the relationship is nevertheless not close.

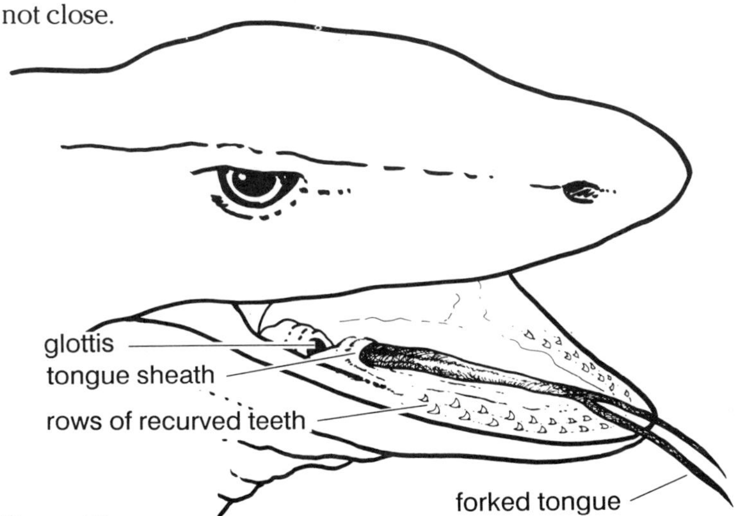

Figure 11
Mouth and tongue of a varanid

The structure of the tongue in varanids is very different from that of other lizards. It is much longer and narrower, with a deep incision in its tip. It does not move freely in the mouth but fits into a sheath when it is retracted (Figure 11). It does not have the roughened dorsal surface found in other lizards. It consists of four parts: a long posterior portion with longitudinal muscle fibres; an area of mainly circular fibres; the main body of the tongue, which has a very complex set of muscles; and the incised tips, which also have muscles running in various directions.

The tongue is partially supported by a complicated structure of bone and cartilage called the hyoid apparatus (Figure 12). The hyoid apparatus of varanids is similar to that of lizards from other families but it has some specialisations, particularly to its anterior portions. There is greater mobility between the joints, and the lingual process enters the connective tissue below the tongue rather than the body of the tongue itself. The superficial throat muscles are exceptionally broad. The hyoid apparatus is relatively heavier and more elongated than in other lizards. This specialisation of the hyoid apparatus, its mobile joints, and associated throat muscles are used to force food into the throat. This has enabled the function of the tongue to change from a manipulatory to a sensory role.

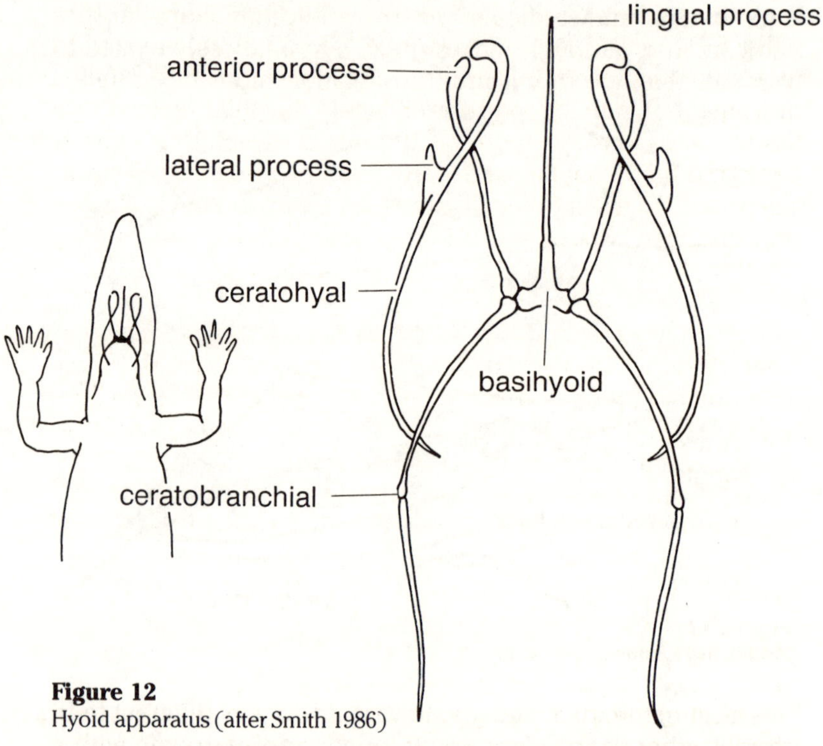

Figure 12
Hyoid apparatus (after Smith 1986)

In its sensory role, the tongue is protruded and flickered rapidly. The tips move independently, thereby increasing their sampling area (Plate 7). Particles are transported to the roof of the mouth close to the openings to the Jacobson's organs. The deeply incised tongue does not insert particles directly into the Jacobson's organs. The function of the long tips is not known but may be to increase their surface area. The tongue is not flattened when it is protruded. It is a cylindrical mass of muscle with a constant volume. When protruded, it becomes longer and thinner. When the tongues of goannas are extended, their length may increase by 72–90 percent. This is much greater than for lizards of other families. Long narrow tongues can be protruded further than shorter, stout ones, and therefore are able to sample a larger volume of air and collect more scent particles than shorter ones. The flickering tongue is highly characteristic of varanids as they search for food.

In most animals, food is transported through the mouth by regular cycles of movement of the tongue, in conjunction with associated jaw movement. Varanids, however, do not use the tongue in the movement of food. Instead, inertial feeding is used, in which the prey is thrown backwards in the mouth and the jaws move rapidly forwards to surround it. Two or three such thrusts may be required to deal with small items of prey while 5–12 may be needed for larger prey. The tongue is protected in its sheath on the floor of the mouth during these movements. Once the food is well within the gular region (soft tissues of the throat; Figure 1), the hyoid apparatus moves it back into the oesophagus. Meanwhile, the neck is bent from side to side to assist in moving the prey down the throat. Large prey are often pressed against other objects to force them into the throat. In captivity, the sides of cages are often used for this purpose.

The tongue is probably not used in transporting liquid when varanids drink. They insert their snout into the liquid and may immerse it entirely. Drinking is probably accomplished by suction created by pumping of the hyoid apparatus and the muscles attached to it. Once liquid reaches the throat, it is forced further back by lifting the head and compressing the hyoid region. This behaviour has been observed in the wild in Komodo dragons (*Varanus komodoensis*), and in several other species in captivity.

Teeth and jaws

The teeth of varanids are large, sharp, and recurved. They are laterally compressed with anterior and posterior cutting edges and are arranged in long tooth-rows (Figure 11). Recurved teeth have greater prey-holding ability than do straight teeth and are useful in catching and successfully handling prey. The teeth are normally hidden in a fold in the gums and are not readily seen. However, their presence is obvious when the animals bite, as they can cause deep lacerations. Handling large prey places great strains on teeth and they often break off where they are attached to the side of the jaw. Replacement teeth develop from the pulp cavity and become fused to the outside of the jaw after bone at the site of the old tooth is resorbed.

The jaw and throat musculature in the larger species of varanids is adapted to feeding on large prey. The jaws are able to close rapidly and are thus suited to the capture of fast-moving prey. After capture, prey must be subdued and correctly orientated before being swallowed. Large items may be given a number of bites, be shaken in the jaws or pressed against objects in the process of being suffocated. Larger prey are smashed against the ground or other solid surfaces. The prey are held securely until all movement ceases. Small prey are subsequently swallowed whole, head first. The larger species of goannas dismember particularly large prey. They hold the prey with their forefeet and tear them into pieces which they can swallow with jerking movements of their heads. In this way, a 50 kilogram Komodo dragon has been seen to dismember and completely devour a 31 kilogram pig carcass in 17 minutes (Plate 4).

In most lizards, the upper jaw can move independently of the base of the skull. This process is called cranial kinesis and is important in assisting the movement of prey into and within the mouth. Compared with its development in other lizards, kinesis is highly developed in varanids. It aligns the upper and lower tooth rows so that prey may be more firmly gripped. This is important when capturing and holding the small elusive prey upon which most species of goannas predominantly feed, and which they are very adept at seizing. Kinesis also allows the mouth to open wider when larger items of prey are being eaten. The part it plays in inertial feeding is also important. When the head is jerked sharply backwards, the jaws move forwards and thus gain a grip further along the body of the prey. The hyoid apparatus (Figure 12) then helps to force the prey into the throat.

From the mouth the food is forced along the oesophagus and into the elongated, muscular stomach (Figure 13) which serves mainly as a storage organ. Beyond the stomach, the pylorus leads into the small intestine (ileum) where most digestion occurs. The ileum in turn leads into the rectum, which is separated from the cloacal chambers by a sphincter muscle. The gut of varanids and other carnivorous reptiles is short compared with those of herbivorous lizards, as digestion is a comparatively simple process in carnivores.

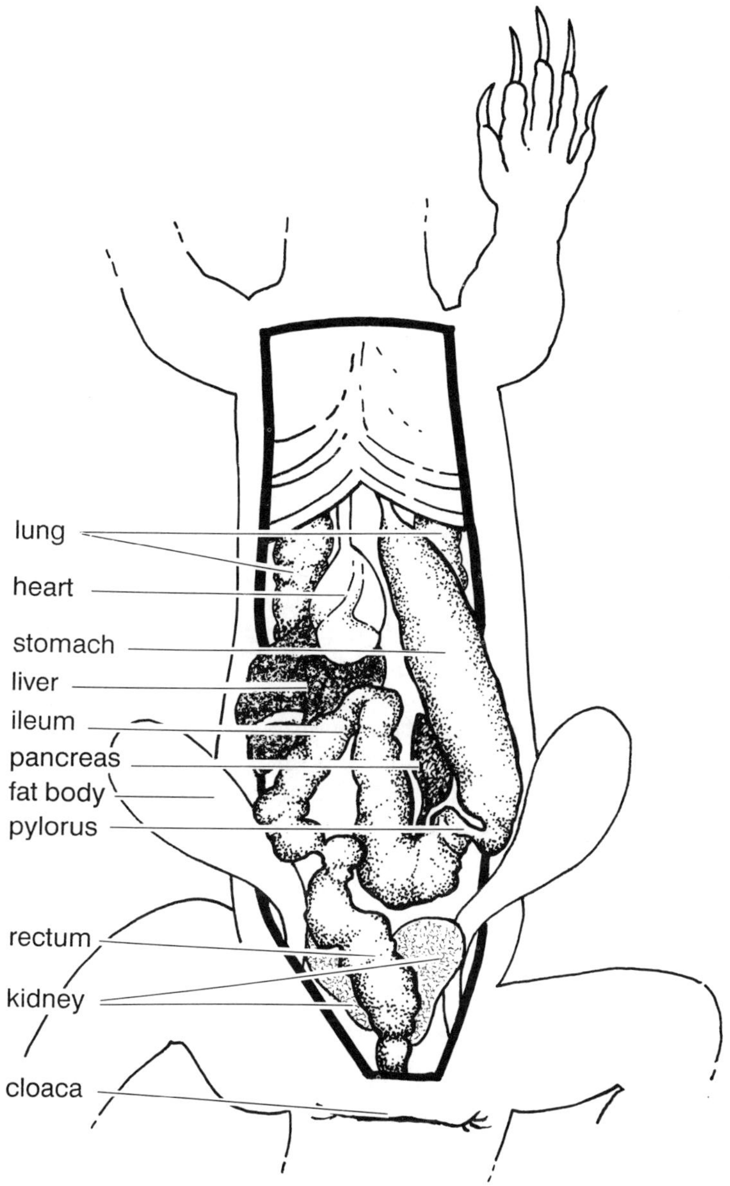

Figure 13
Internal organs of *V. rosenbergi*

CHAPTER

4

BREEDING

SEX ORGANS

Male goannas are distinguished by their generally larger size and the presence of two lateral bulges at the base of the tail. Each bulge contains a hemipenis, one of the copulatory organs of the male. Each hemipenis is everted from the cloacal opening whenever the animal is aroused, either for copulation or in response to threat. The hemipenes are erected and retracted by means of muscular contractions and variations in blood supply. These organs are very elaborate structures (Figure 8). The flamboyant appearance of the hemipenes is probably important during courtship display or mating. The hemipenes of varanids each have two lobes, and each lobe bears a terminal cartilage or bone, called the hemibaculum.

Female Rosenberg's goannas also have organs that resemble hemipenes and although smaller and less developed than those of males, these can often make it very difficult for humans to determine the sex of this species.

The testes and ovaries of varanids are located within the abdomen (Figure 14) and are relatively small when the animals emerge from the general inactivity of winter. During spring the gonads of both sexes start to increase in size (Figure 15) and gametes (sperm and ova) begin to develop. The testes reach their maximum size in December and at about the same time the ovaries shed ova into the oviducts. Over several weeks the ova increase substantially in size as large quantities of creamy coloured yolk are added to them. By mid-summer Rosenberg's goannas are ready to mate.

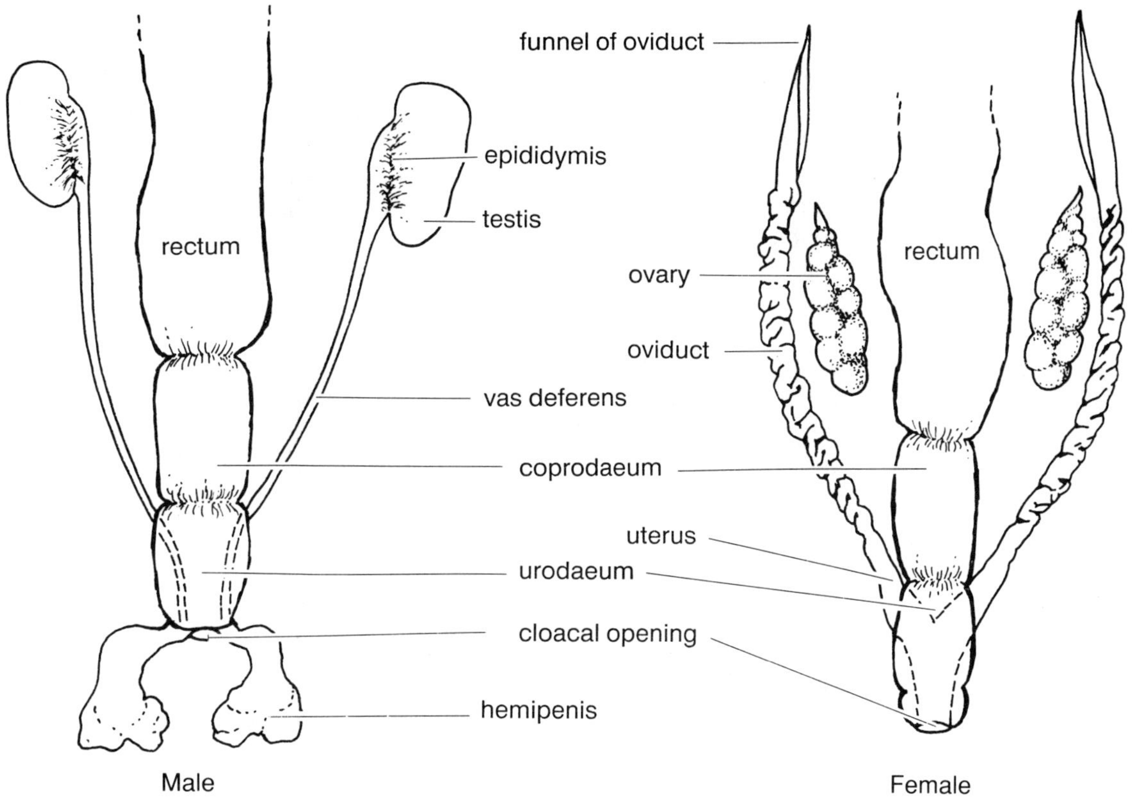

Figure 14
Sex organs of a male and female varanid

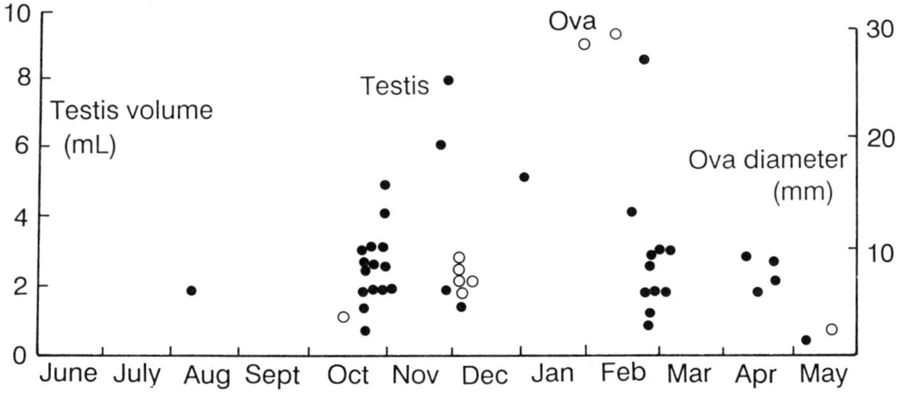

Figure 15
Seasonal changes in size of testes and ova in *V. rosenbergi*

MATING

Mating in *Varanus rosenbergi* is a surprisingly tender event. Once a male goanna has located the burrow of a female that is in breeding condition, he constructs a burrow of his own a few metres away. Over several days the male and female spend an increasing amount of time in each other's company. Eventually the pair begin to copulate. Copulation is preceded by the male entering the burrow of the female, leaving his tail protruding from the burrow entrance. After a few minutes he backs slowly out of the burrow, leading the female, head first, to the surface. As he leads her, he nuzzles and licks her neck, and when they are both in the open outside the burrow he frequently rubs the side of his head along her back. He also nuzzles her cloacal opening before moving his body forward across her back, until their cloacas are side by side. He may approach her from either side as he has paired copulatory organs. The male uses his tail to raise that of the female off the ground, while the female raises her hind leg closest to him, allowing him to insert the adjacent hemipenis. Insertion is accompanied by a few lateral pelvic thrusts after which the pair remain coupled and still for about ten minutes (Figure 16). The female, who appears extremely nervous at this time, then breaks free and retires to her burrow.

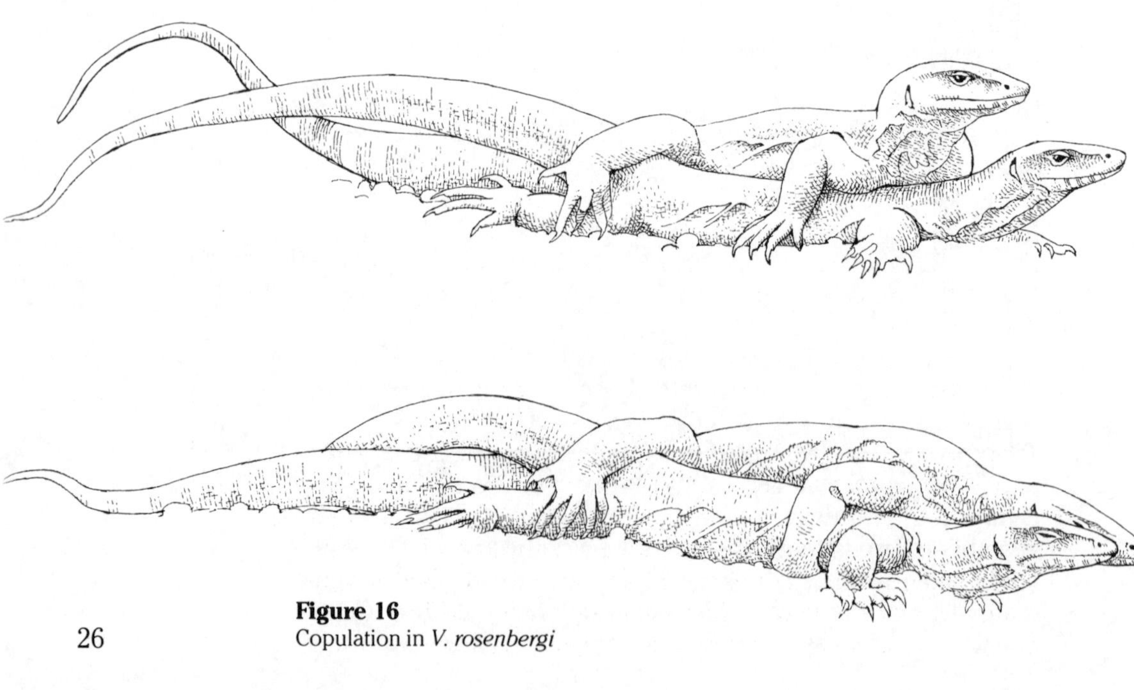

Figure 16
Copulation in *V. rosenbergi*

The male remains on the surface, basking on the entrance mound and occasionally digging and scraping further soil onto it. After a further ten minutes or so, the male re-enters the female's burrow and persuades her to re-emerge for further copulations. This activity goes on for several days, with the male using his hemipenes alternately. During this period of intense breeding activity the pair may come to share the same burrow. After some days, the intensity of copulation declines and finally only occupies a few hours in the morning. After this, the goannas separate and forage independently.

If another male enters the immediate vicinity of the breeding pair, the couple separates and the female retreats to her burrow while the male chases off the intruder.

The mating behaviour of several other species of varanids has been observed in the wild (*Varanus komodoensis* and *Varanus varius*) and in captivity (*Varanus bengalensis, Varanus olivaceus, Varanus timorensis* and *Varanus gilleni*). The behaviour of all species is similar but there are some characteristic differences between species. Scratching, nuzzling the head and neck area and licking are important in all species. In some (*Varanus komodoensis, Varanus olivaceus* and *Varanus gilleni*) the male occasionally bites the female while he lies on her back, but this was not seen in *Varanus timorensis* or *Varanus rosenbergi*. In *Varanus bengalensis* the female sometimes bites the male during mating. In others (*Varanus komodoensis* and *Varanus bengalensis*) the male immobilises the female during mating by holding her in a tight embrace with his front legs. Females of these species are aggressive during mating, and could damage the males by biting or clawing them. This immobilisation has not been observed in *Varanus varius, Varanus rosenbergi* or *Varanus olivaceus*. In *Varanus olivaceus, Varanus timorensis* and *Varanus varius* the males use either hindleg to lift the base of the female's tail prior to inserting the hemipenis. The mating behaviour of male varanids of different species has to be flexible so they can react to varied responses of the females.

Varanids generally copulate frequently over a period of several days. One pair of wild *Varanus varius* was seen to copulate 16 times in three hours.

Some species (*Varanus komodoensis, Varanus bengalensis* and *Varanus rosenbergi*) seem to form pair bonds of varying duration. In captivity mating occurs only after both animals have been kept in the same cage for a lengthy period. However, wild female *Varanus varius* have been seen mating with several males within a few days. Mating assemblages, which usually consist of one female and several males, have also been seen

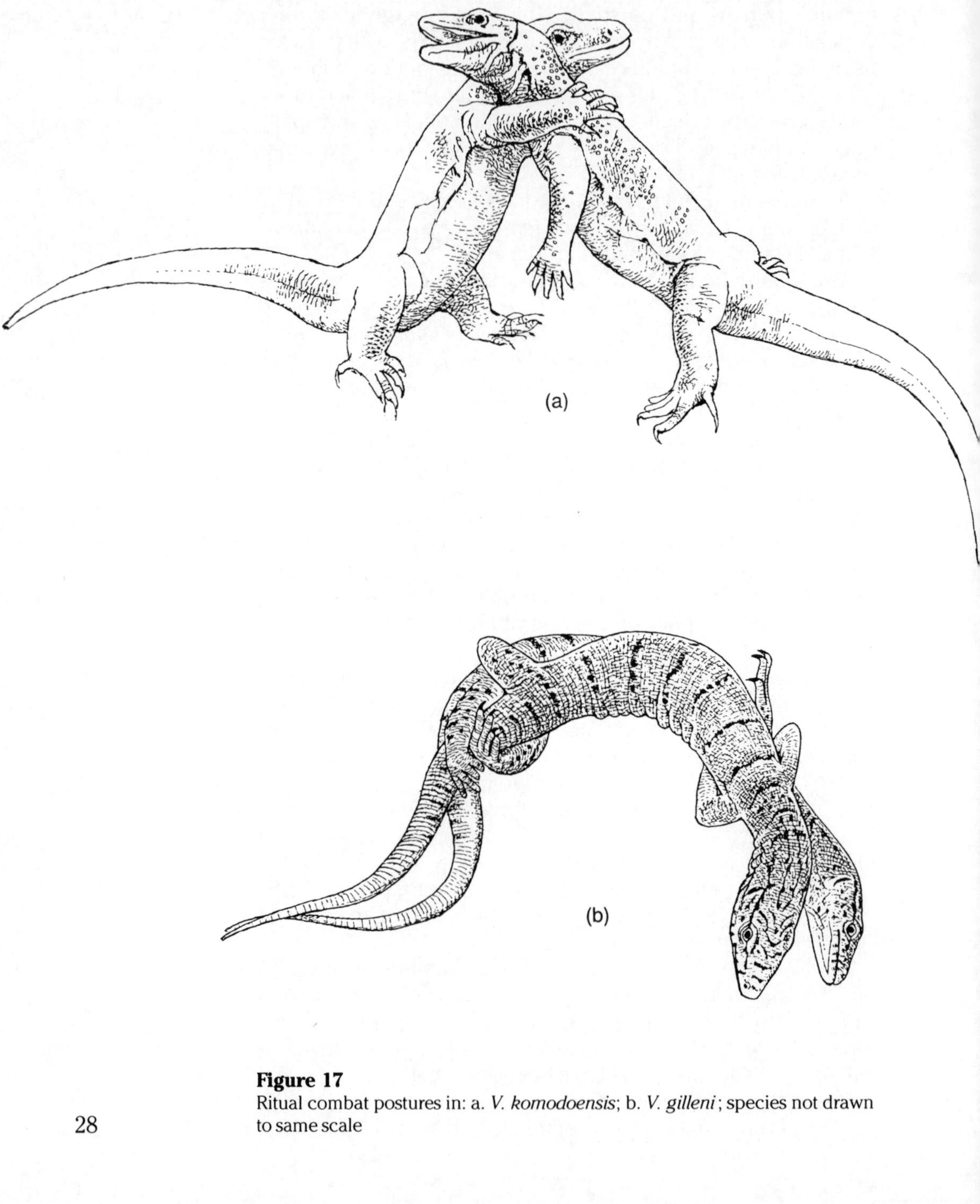

Figure 17
Ritual combat postures in: a. *V. komodoensis*; b. *V. gilleni*; species not drawn to same scale

in *Varanus varius*. The largest male eventually chased away other males which approached the female.

Some male varanids engage in what is called ritual combat. The larger species (*Varanus komodoensis, Varanus bengalensis, Varanus olivaceus, Varanus salvator, Varanus spenceri, Varanus varius, Varanus gouldii* and *Varanus mertensi*) initially wrestle one another in an upright stance and clutch each other with the forelegs only (Plate 13), whereas the smaller species (*Varanus gilleni, Varanus semiremex, Varanus caudolineatus* and *Varanus timorensis*) clutch one another with both the forelegs and the hindlegs during the early stages of these encounters. Such fights can also involve the antagonists rolling along the ground while entwined (Figure 17). During ritual combat varanids try to overpower one other, and severe wounds are occasionally inflicted by the teeth and claws. Usually it is only after dominance is determined that the subordinate animal is bitten by the winner. Some species (*Varanus bengalensis* and *Varanus salvator*) apparently do not bite one another during this combat, while other species do (*Varanus gilleni, Varanus niloticus, Varanus olivaceus, Varanus komodoensis* and *Varanus varius*).

EGG-LAYING

The time between fertilisation and egg laying is not known, but during this period the leathery shells that are characteristic of many reptiles are formed in the lower portion of the oviducts. After this process occurs, the female *Varanus rosenbergi* locates an active termite mound (usually *Nasutitermes* species) and begins to excavate an access hole at the top, or high on one side of the mound. She digs a tunnel approximately 50–60 centimetres deep, towards the centre of the termitarium. At the end of this shaft she makes a large cavity (Figure 18). The digging of this egg chamber is usually completed within a day, but can take longer if the termite mound is very hard. On completing the excavation the female sits in the access shaft and begins to lay her clutch of eggs into the deep chamber, with only her head protruding from the mound. Usually 10–17 eggs are laid, weighing approximately 26 grams each and measuring about 50 x 30 millimetres. The eggs of other varanids differ greatly in size. The average weight of those of *Varanus gilleni* is only 4 grams and they measure 16 x 27 millimetres, while those of *Varanus komodoensis* weigh 125 grams and are 55 x 85 millimetres. Varanid eggs are extremely rich in lipids (fats) that provide the main energy source for the developing embryo. The general composition of goanna eggs is shown in Table 3.

Egg-laying occupies a few hours, after which the female begins to refill the excavation. She pushes soil up the mound using her hindlegs, scratching surface mound material towards the hole with her forelimbs. The female takes several rests during the refilling process. She often moves around the summit region of the mound, apparently marking it with scent from her cloacal region. She may also stop to bask. The male may

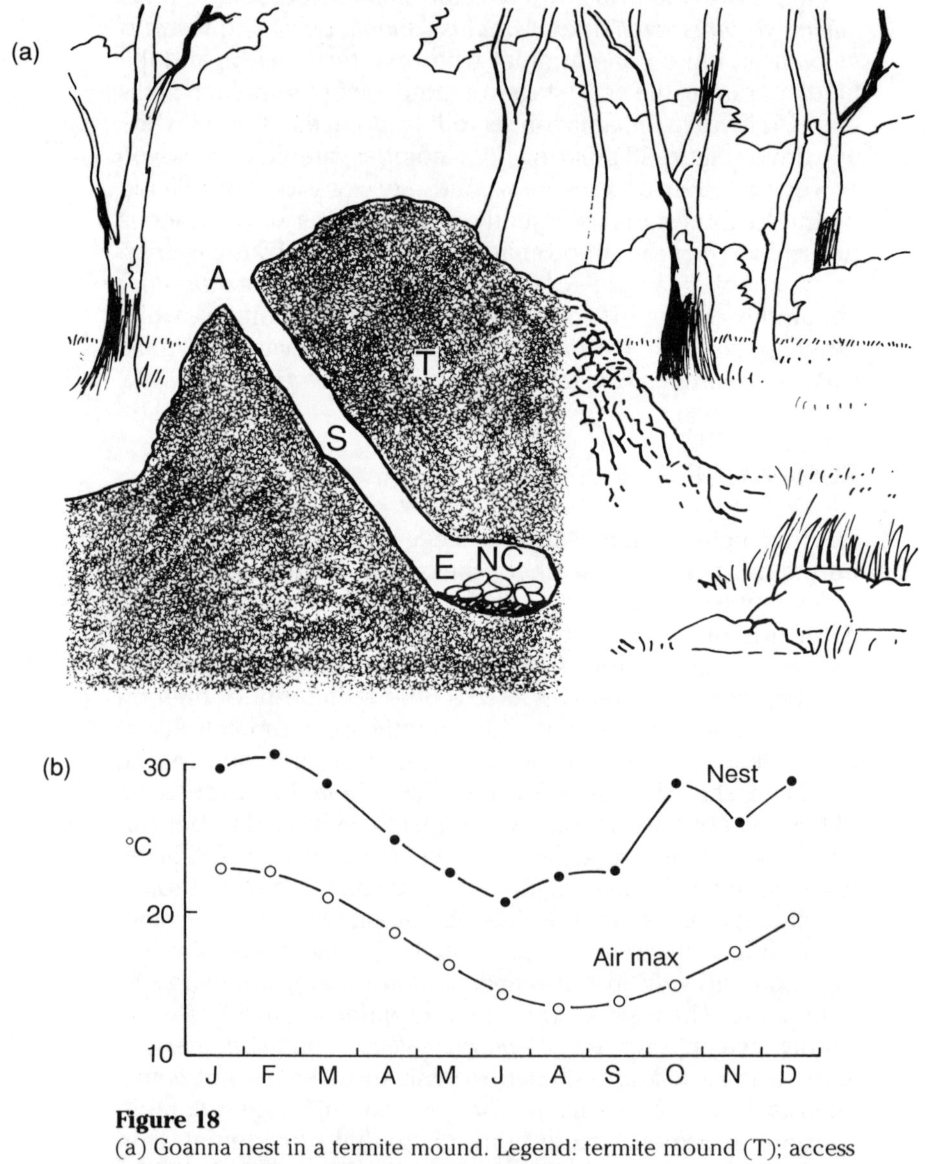

Figure 18
(a) Goanna nest in a termite mound. Legend: termite mound (T); access hole (A); shaft (S); nest chamber (NC); eggs (E)
(b) Monthly changes in nest and air temperature

Table 3
Composition of three eggs of *Varanus rosenbergi*

Egg	Weight (g)	Dry matter (%)	Water (%)	Shell (%)	Fat (%)	Protein (%)	Energy (kJ/g)
1	25.69	32.0	68.0	2.64	13.41	11.55	8.17
2	24.88	33.3	66.7	2.88	14.30	11.72	8.56
3	26.39	28.9	71.1	2.87	13.75	8.99	7.69
Mean	25.65	31.4	68.6	2.80	13.82	10.75	8.14

associate with the female as she labours to completely close off the egg chamber, but he performs only occasional desultory scratching at the mound before returning to his foraging. Both adults pay regular visits to the termitarium in the days immediately after egg-laying. This probably prevents other goannas from attempting to deposit their eggs in the same mound. It may also serve to ward off any diurnal predators trying to rob the nest chamber before the termites have reconstructed the mound. During this post-mating period the male may still attempt to court and mount the female. She rejects these attempts by adopting a raised aggressive stance. Egg laying is generally completed by late summer in *Varanus rosenbergi*, after which the mating pairs separate. The gonads decline in size at the end of the breeding season (Figure 15).

The termites reconstruct the mound around the goanna eggs so that they become totally encapsulated. Termitaria are excellent incubation chambers for reptile eggs. The termites regulate the temperature of the mound to around 30°C throughout most of the year, apart from a slight cooling in winter. The humidity in the mound is always near saturation. Therefore, the goanna eggs and embryos are maintained at a suitable temperature and humidity during their development (Figure 18). The eggs are able to absorb moisture from their surroundings and increase in weight as they develop.

Although not all species of goannas lay their eggs in termitaria, at least some of the larger species are known to do so (*Varanus albigularis, Varanus bengalensis, Varanus giganteus, Varanus gouldii, Varanus niloticus, Varanus rosenbergi* and *Varanus varius*). The surfaces of the termite mounds are very hard, and only large species with sharp and powerful claws can open the mounds and use them for nesting places. Termitaria provide protection for the eggs from predators. Some species (*Varanus panoptes* and *Varanus bengalensis*) bury their eggs deep in the soil, especially along the margins of creek beds. Also, some varanids appear to lay their eggs in communal nests that give the appearance of a warren (*Varanus komodoensis*, *Varanus bengalensis* and *Varanus panoptes*). The locations of the nests of small species are not known.

CLUTCH SIZE

The clutch size of varanids is generally related to their body size (Figure 19a and b). However, there are some exceptions. Some species, such as the desert-dwelling *Varanus tristis*, produce uncharacteristically large clutches of small eggs. The maximum clutch size reported in varanids is 46 eggs in *Varanus exanthematicus*.

Fully developed eggs occupy all the available space in the body cavity of female varanids and the animal becomes highly distended. In at least two species (*Varanus tristis* and *Varanus scalaris*) most of the gravid females carrying nearly developed eggs have empty stomachs, probably because of the limited abdominal space remaining and the pressure exerted on the internal organs by the eggs. Feeding probably stops during the late stages of egg development in most varanid species. Female *Varanus exanthematicus* are reported to be largely inactive for approximately 1 month prior to egg-laying.

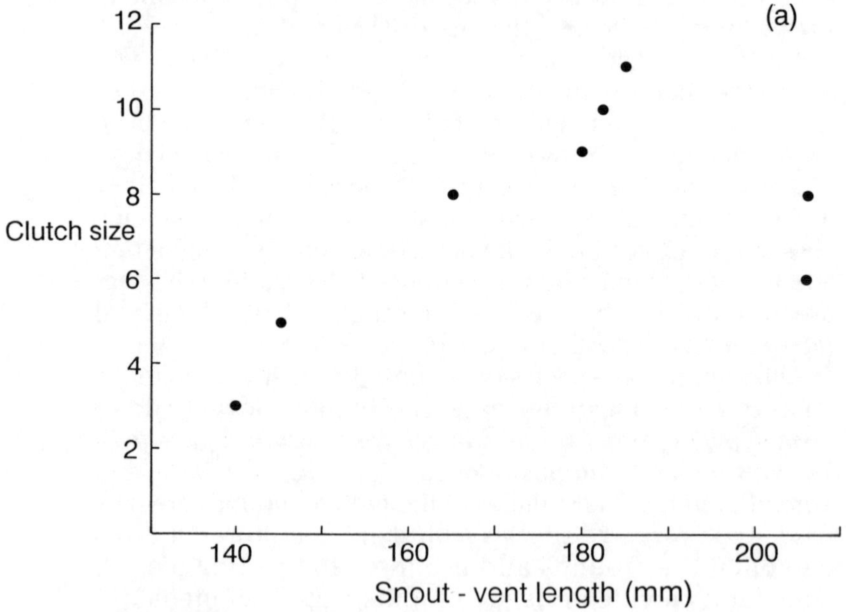

Figure 19(a)
Clutch size and body size in *V. acanthurus*

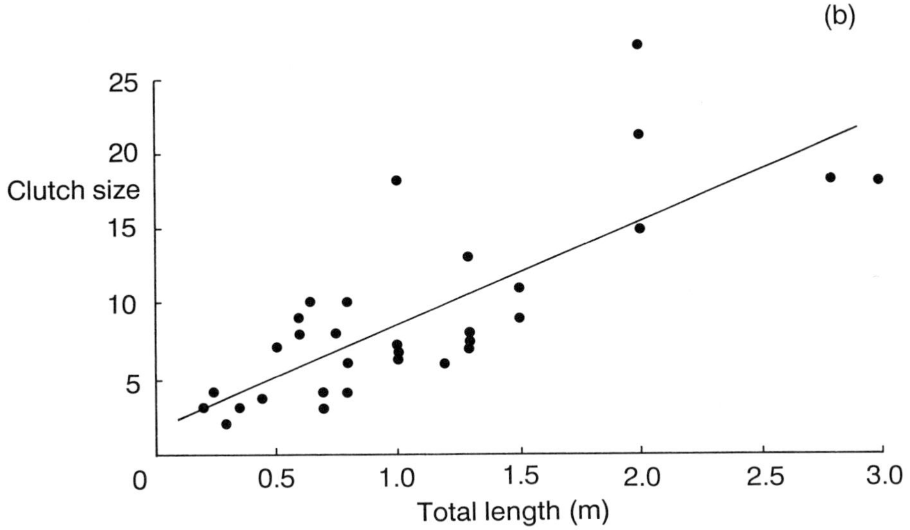

Figure 19(b)
Clutch size and body size in varanids

Data obtained from eggs of captive animals, which have been incubated at similar, constant temperatures, show that incubation time is also related to body size. The eggs of larger species take longer to hatch, with incubation periods ranging from 77 days for *Varanus brevicauda* to 250 days for *Varanus salvator* (Figure 19c). Incubation periods under constant temperatures in captivity are shorter than those observed in the wild. This may be due to low and fluctuating temperatures which clutches experience in natural situations. It may also indicate the delayed emergence of young from nests in termitaria.

The timing of breeding of different species varies considerably. In areas of northern Australia, some species breed in the wet season while others in the same region breed in the dry season. The onset of breeding, even in those which breed in the same season, may differ by several months. Populations of the same species inhabiting different climatic regions may start breeding several months apart. More than one factor may determine the time of breeding. Females require an adequate food supply to be able to produce eggs. Soil conditions must be suitable to enable the female to dig a nest that provides a suitable microclimate for the eggs. Hatchlings need a good food supply to survive and grow rapidly.

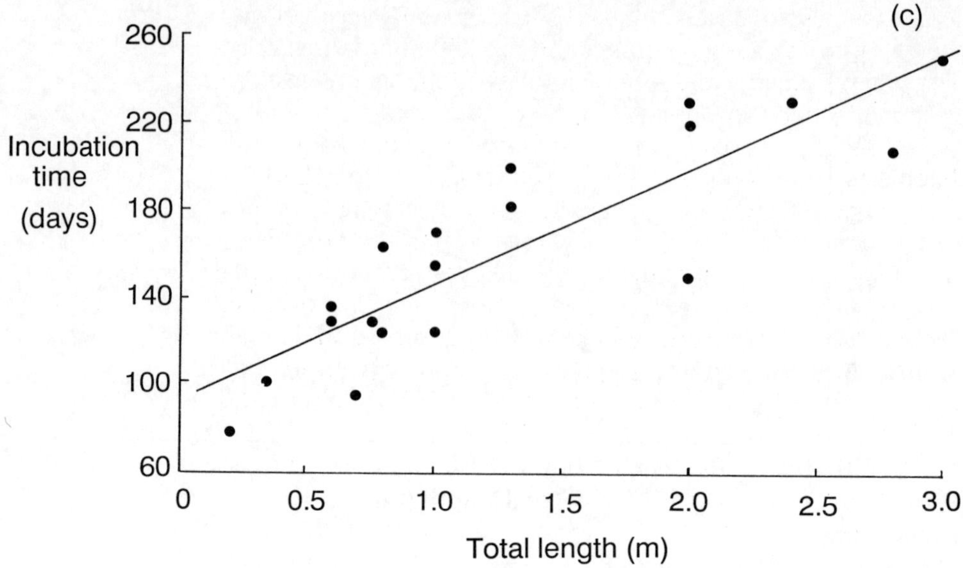

Figure 19(c)
Incubation time and body size in varanids

HATCHING

The eggs of *Varanus rosenbergi* and *Varanus varius* incubate in termitaria over winter and hatch in early spring, about eight months after they are laid. It is not certain whether the hatchlings are capable of digging their own way out of the termite mound, or if they rely on assistance from adults. The construction material of some termite mounds is exceedingly hard and would present major difficulties for hatchlings trying to emerge. One clutch of *Varanus rosenbergi* eggs was accidentally disturbed during farming operations in late September. Hatchlings were found still cocooned within the termitarium although they had escaped from their egg shells.

Hatchlings of other species emerge at various different times of the year, when conditions are favourable. Those of *Varanus acanthurus*, and probably those of other pygmy goannas, appear in early summer after approximately 2½ months of incubation. In eastern Indonesia, hatchlings of *Varanus komodoensis* emerge at the start of the dry season, while those of *Varanus timorensis* apparently emerge at the end of the dry season.

Hatchlings of different species vary greatly in size. *Varanus rosenbergi* hatchlings weigh about 18 grams and are approximately 170 millimetres long. Those of the pygmy species *Varanus gilleni* and *Varanus storri* are much smaller, approximately 2 grams in weight and 120 millimetres long. The hatchlings of larger species are much larger. Those of *Varanus komodoensis* weigh approximately 80 grams and are 300 millimetres long. The relative lengths of the body and tail in the hatchlings also vary greatly between species.

The hatchlings of *Varanus rosenbergi* are very active and adept at climbing, using their prehensile-like tail. Little is known about the general behaviour of newly hatched *Varanus rosenbergi* (Plate 8) as they are rarely seen in the field. They may spend most of their time in trees, like the juveniles of other species such as *Varanus komodoensis, Varanus bengalensis* and *Varanus varius*. They are less likely to be observed there than animals that are active on the ground. Varanids are cannibalistic and they and other predators could readily capture and eat hatchlings of all varanid species.

Varanid hatchlings are generally much more colourful and strongly patterned than the adults. The hatchlings of *Varanus rosenbergi* (Plate 9) are black, white and grey with pale yellow markings and their throats and heads have bright orange patches. The hatchlings of *Varanus glauerti* are a bright iridescent blue, while those of *Varanus dumerili* (Plate 10) are bright yellow and black, with an orange head. The reasons for the bright colouration of the young are unknown, but it may serve either as camouflage or as a deterrent, for many toxic species have very bright colouration which acts as a warning to predators.

GROWTH RATES

No data are available on the growth rates of young *Varanus rosenbergi*. Small individuals are seldom captured in the field and none have been bred successfully in captivity. The early growth rate data that are available for a few other species of goannas are almost totally limited to those from captive animals. Growth rates are probably influenced by the conditions under which the animals were kept, particularly by the amount of food provided. In most cases, the young captive goannas probably received more food than would have been available to wild animals.

The duration of growth seems to be influenced by the adult size of the species. The young of all species rapidly increase their weight (by 2–4 times) and their total length (by 1.2–2 times) during their first 3 months of life. After one year the smaller species are approaching their maximum length (approximately twice that at hatching), whereas the length of the larger species, such as *Varanus varius* (5–6 times hatchling length) and *Varanus komodoensis* (up to 10 times hatchling length) continues to increase until they are well over five years old (Plate 11).

We also do not know at what age *Varanus rosenbergi* starts to breed. The estimated age of sexual maturity for *Varanus komodoensis*, based on data from captive animals, is approximately 5 years and that for *Varanus bengalensis* is 3 years for males and 4 years for females.

LONGEVITY

Few data are available on the longevity of varanid lizards, as no long-term field studies have been carried out to determine the population dynamics and survival rates of any species. The longest interval between the first and last capture dates of an individually marked *Varanus rosenbergi* on Kangaroo Island was 74 months. When first captured, it was a 1290 gram adult with a 420 millimetre snout–vent length. It did not increase in length during the following 74 months and its weight only increased to 1320 grams. No data are available on the maximum survival time of *Varanus rosenbergi* in captivity. The longest time between captures of a wild *Varanus komodoensis* is 8.5 years.

At least nine species of varanids are known to have lived for more than five years in captivity. Records of the longest periods during which species have been kept in captivity are shown in Table 4.

Table 4
Longest known periods in captivity of varanids

Species	Period
V. bengalensis	5–6 years
V. exanthematicus	17+ years
V. flavescens	10 years 6 months
V. gouldii	approx. 7 years
V. griseus	9 years 6 months
V. komodoensis	16 years 7 months
V. niloticus	15 years
V. salvator	10 years 8 months

Note: Age of animals was unknown at beginning of captive period

CHAPTER 5

GENERAL BEHAVIOUR

HOME RANGE

The home range of an animal is the total area used for all of its activities. Little is known about the sizes of the home ranges or activity areas of goannas, although accounts of their activity levels generally state that most species are wide-ranging, active foragers.

The home ranges of 13 *Varanus rosenbergi* were determined on Kangaroo Island by recapturing animals over periods ranging from 5 to 74 months. The capture locations were plotted on a map, and the outermost points were connected and the areas within were measured (Figure 20). The home range areas (Table 5) were highly variable, as has been found in the limited data on home ranges for other varanids (*Varanus giganteus*, *Varanus komodoensis* and *Varanus griseus*). The reasons for this variability are unknown, but it may be due to behavioural differences between the sexes, or to variations in hunting ability and availability of prey. The sex of most species of goannas, including *Varanus rosenbergi*, is very difficult to determine with certainty on the basis of external morphology. Therefore it was not possible to relate size of the home ranges of goannas on Kangaroo Island to their sex. The home ranges of some *Varanus rosenbergi* overlapped to a considerable extent (Figure 20).

There is an empirically derived equation which predicts the size of the home range of a lizard on the basis of its body weight. The values determined for the home ranges of *Varanus rosenbergi* are greater than those predicted by this method. However, the predicted home ranges are based mainly on data obtained from herbivorous and insectivorous species of lizards, which are thought to use smaller areas than carnivorous species such as the large varanids.

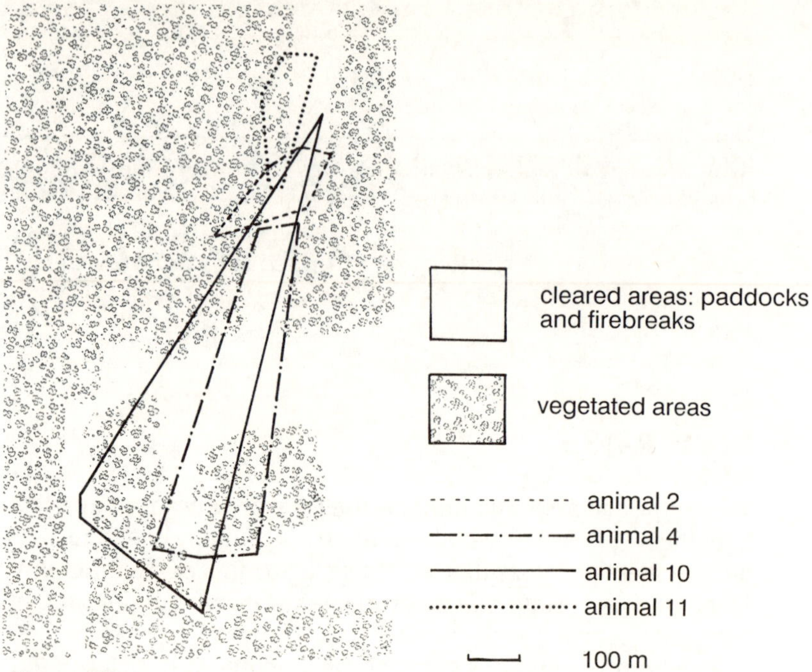

Figure 20
Home ranges of *V. rosenbergi*

Table 5
Home range sizes of *Varanus rosenbergi* on Kangaroo Island

Animal number	Weight at last capture (g)	Months from first to last capture	Home range (ha)
1	770	10	7.22
2	990	24	4.14
3	1060	18	36.95
4	1370	28	37.52
5	1795	25	1.71
6	735	5	5.57
7	1350	32	14.21
8	1320	74	16.19
9	950	41	6.18
10	1465	46	43.70
11	700	32	3.36
12	1425	55	36.92
13	1010	58	38.99
Mean	1149		19.44

ACTIVITY AREAS

An activity area is that portion of a home range used during a period such as a day or a season. On Kangaroo Island, *Varanus rosenbergi* were active away from their burrows on most suitable days throughout the year. The sizes of daily activity areas of goannas fitted with radio-transmitters were determined by plotting their locations several times each day. The activity areas (which were calculated in the same way as home ranges) varied seasonally, the largest being used during summer and the smallest during winter (Table 6). During the warmer months, the animals moved furthest, as goannas then reached their highest body temperatures and were active for the longest periods each day. In the warmer months, goannas were commonly found on low-lying, cleared areas. During winter, when these areas become very wet, the goannas moved onto elevated, well-drained and more densely vegetated sites.

Table 6
Seasonal variation in the mean size of daily activity areas (in hectares) of *V. rosenbergi* on Kangaroo Island

Month	Days of observation	Daily activity area (ha)
June	10	0.18 (0.16–0.20)
August	14	0.71 (0.19–1.11)
December	17	1.37 (0.67–2.43)
March	17	0.66 (0.62–0.70)

Values in brackets are the range.

ACTIVITY PATTERNS

The seasonal activity patterns of different species of goannas can differ greatly. Many tropical species are active throughout the year, whereas others, particularly those living in temperate or desert regions, may be inactive during the winter or in drought conditions.

The daily activity pattern of goannas in temperate or tropical areas normally consists of a single (or unimodal) period of activity during the day in all seasons. The lizards emerge from their overnight shelter in the morning and remain outside until the late afternoon or evening and then retreat back into shelter. On Kangaroo Island the activity period is longer during the summer than in the cooler months. Desert dwelling species generally have two periods of activity (bimodal) during the day in the hotter months. Desert forms emerge earlier in the day than in cooler seasons and retreat to their burrow, or into some other shelter site, in the middle of the day when it is hottest. Some goannas emerge again for a second period of activity during the afternoon. During the cooler months they may adopt a unimodal activity pattern or remain inactive in their burrows. They may not emerge from their burrows at all during winter, perhaps because of a lack of available prey, or because low soil or burrow temperatures do not provide the stimulus for them to emerge.

In many species of goannas and other lizards, males have larger territories and are more active within them than are females. As most goannas are encountered by chance when they are active, those individuals which are most active will be those most frequently encountered. Sex ratios in museum collections of many species of varanids are strongly biased in favour of males (*Varanus eremius, Varanus glauerti, Varanus glebopalma, Varanus semiremex, Varanus caudolineatus, Varanus storri, Varanus tristis, Varanus scalaris, Varanus gouldii, Varanus panoptes, Varanus mertensi, Varanus mitchelli, Varanus indicus, Varanus varius, Varanus komodoensis, Varanus salvator* and *Varanus rosenbergi*). A sample of 45 dead *Varanus rosenbergi* from Kangaroo Island consisted of 37 males and 8 females, a ratio of 4.6:1, which was probably due to differences in the levels of activity between the sexes. However, because of the difficulty in determining the sex of live goannas of this species, it is not definitely known whether these behavioural differences occur between male and female *Varanus rosenbergi*. Another species, *Varanus acanthurus*, is generally captured at its shelter sites. There is no male bias in museum collections of this species, since there are no differences in activity levels between the sexes to influence the sex ratio of captured animals.

BURROW USE

The entrance of a goanna burrow is often situated beneath a flat rock, a small shrub or a fallen log (Figure 21). There is usually a mound of earth at the burrow entrance and the animal often lies upon this to bask for some time after emerging from the burrow. The burrows of *Varanus gouldii* often have a small

Plate 1: Rosenberg's monitor (*V. rosenbergi*) from Kangaroo Island, South Australia. They are the largest terrestrial predator on the island and are very abundant there. (Photograph by Brian Green)

Plate 2: The Komodo dragon, *V. komodoensis*, from several islands in eastern Indonesia, is the largest living species of lizard. (Photograph by Dennis King)

Plate 3: This species is often referred to as *V. 'pellewensis'*, which is not its official name. It lives in northern Australia, and will probably be described as a new species in the near future. (Photograph by John Wombey)

Plate 4: The jaw bones and muscles of Komodo dragons (*V. komodoensis*) are powerful enough to enable them to tear apart and eat large items of prey, such as pigs, deer and buffalo. (Photograph by Dennis King)

Plate 5: The amphibious species *V. mertensi*, mainly hunts in the waters of rivers and billabongs for fish and crustaceans, although it will also forage on land. (Photograph by Graeme Chapman)

Plate 6: The perentie, *V. giganteus*, lives in central Australian deserts. The diet of adults mostly consists of large prey items. (Photograph by Hans-Georg Horn)

Plate 7: The extended tongue of this *V. panoptes* from northern Australia can sample a very large area for scent particles. This assists them in locating food or mates. (Photograph by Harry Butler)

Plate 8: The young of *V. rosenbergi* hatch in the spring, often in termitaria where the eggs incubate over winter at a relatively high and constant temperature. (Photograph by Graeme Chapman)

Plate 9: The reason for the bright colouration of hatchlings of *V. rosenbergi* is unknown, but it may provide camouflage for them as they shelter on the forest floor. (Photograph by Brian Green)

Plate 10: Hatchlings of *V. dumerili* are much more colourful than the adults, as is the case with most goannas. (Photograph by Hans-Georg Horn)

Plate 11: Young *V. komodoensis* grow very rapidly during the first few years of their lives, and reach adult size when they are about five years old. (Photograph by Brian Green)

Plate 12: Adult *V glauerti* are more colourful than most other species of goannas. They live in rocky areas of the Kimberley region of Western Australia. (Photograph by R J (Hank) Jenkins)

Plate 13: Some large species of goannas, including *V. mertensi*, engage in a bipedal form of ritual combat during the mating season. (Photograph by Richard Braithwaite)

Plate 14: An aggressive posture is taken by a *V gouldii* in sand dunes near Lake Eyre. Many species of goannas adopt similar postures when threatened. (Photograph by Dennis King)

Plate 15: Large numbers of the colourful skins of *V. salvator cumingi* are exported each year from the islands of Samar, Leyte, Bohol and Mindanao in the Philippines. (Photograph by Maren Gaulke)

shaft that runs from the terminal chamber of the burrow up towards the soil surface. If the goanna is within the terminal chamber and a predator begins to enter the burrow, the goanna can use this shaft, which is called a 'pop-hole', as an escape route (Figure 21).

In most cases, goannas dig their own burrows and normally live alone in them. However, some *Varanus rosenbergi* occasionally share a burrow with a non-breeding animal of the same species, and at other times a burrow will be occupied by different goannas on different nights. Individual *Varanus rosenbergi* usually frequent more than one burrow, but there is no detailed information on the frequency of use of a burrow or the numbers of individuals that use a particular burrow.

The warrens of introduced rabbits (*Oryctolagus cuniculus*) occur widely within the ranges of *Varanus gouldii, Varanus rosenbergi* and *Varanus varius*. These goannas frequently use rabbit warrens for shelter instead of constructing their own burrows.

Burrows not only play an important role in protecting a goanna from predators. They are also very important in providing protection against the weather (see Thermal Biology, Chapter 6).

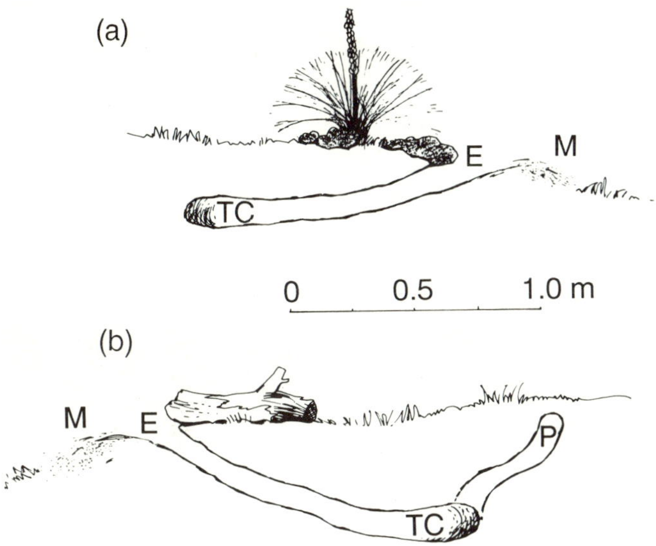

Figure 21
Burrows of (a) *V. rosenbergi*; (b) *V. gouldii*; mound (M); entrance (E); terminal chamber (TC); 'pop-hole' (P)

Figure 22
Tracks of a varanid

MOVEMENT AND BODY POSTURES

When walking, goannas carry their bodies high off the ground and only a small part of the tail near the tip touches the ground surface. This gives rise to the typical goanna tracks seen in sandy or dusty soils (Figure 22). When running, the tail is lifted clear of the ground (Figure 23).

In its usual, non-aggressive posture a varanid has most of its body resting on the ground while the head and shoulders are slightly raised (Figure 24a). When the animal adopts an aggressive stance, it remains on all fours, with the gular pouch and abdomen inflated (Figure 24b and c (Plate 14)). The goanna hisses loudly and can produce powerful side-swipes of the tail (Figure 24b and c). In several species, including *Varanus giganteus*, these side swipes are said to be capable of breaking the legs of dogs. The most aggressive posture in varanids is where the animal stands upright, inflates the gular pouch and makes loud hissing sounds from the nostrils (Figure 24d). The animal may sway slightly. This posture is also used in ritual combat by some species. Aggressive postures are directed not only towards members of the same species. They are also directed at potential predators such as humans, and sometimes at venomous snakes.

A common posture in some varanids is a bipedal stance in which the animal raises itself vertically and supports itself on its hind legs and tail (Figure 25). The animal then has a clear and elevated view of its surroundings, and this posture may have given rise to the 'monitor' name.

The behaviour patterns associated with hunting and breeding are discussed in Feeding (Chapter 3) and Breeding (Chapter 4).

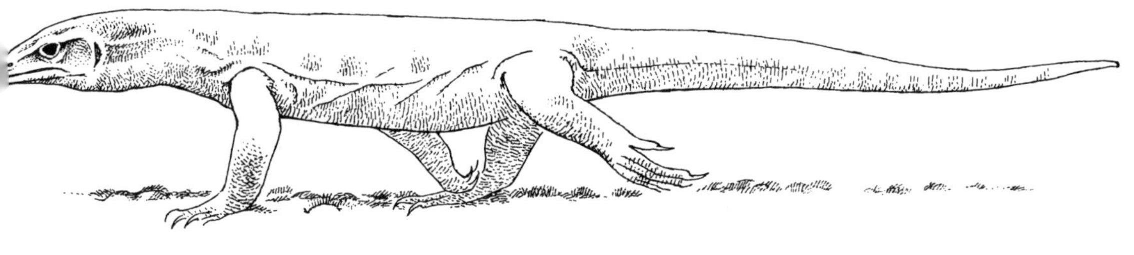

Figure 23
Running *V. komodoensis*

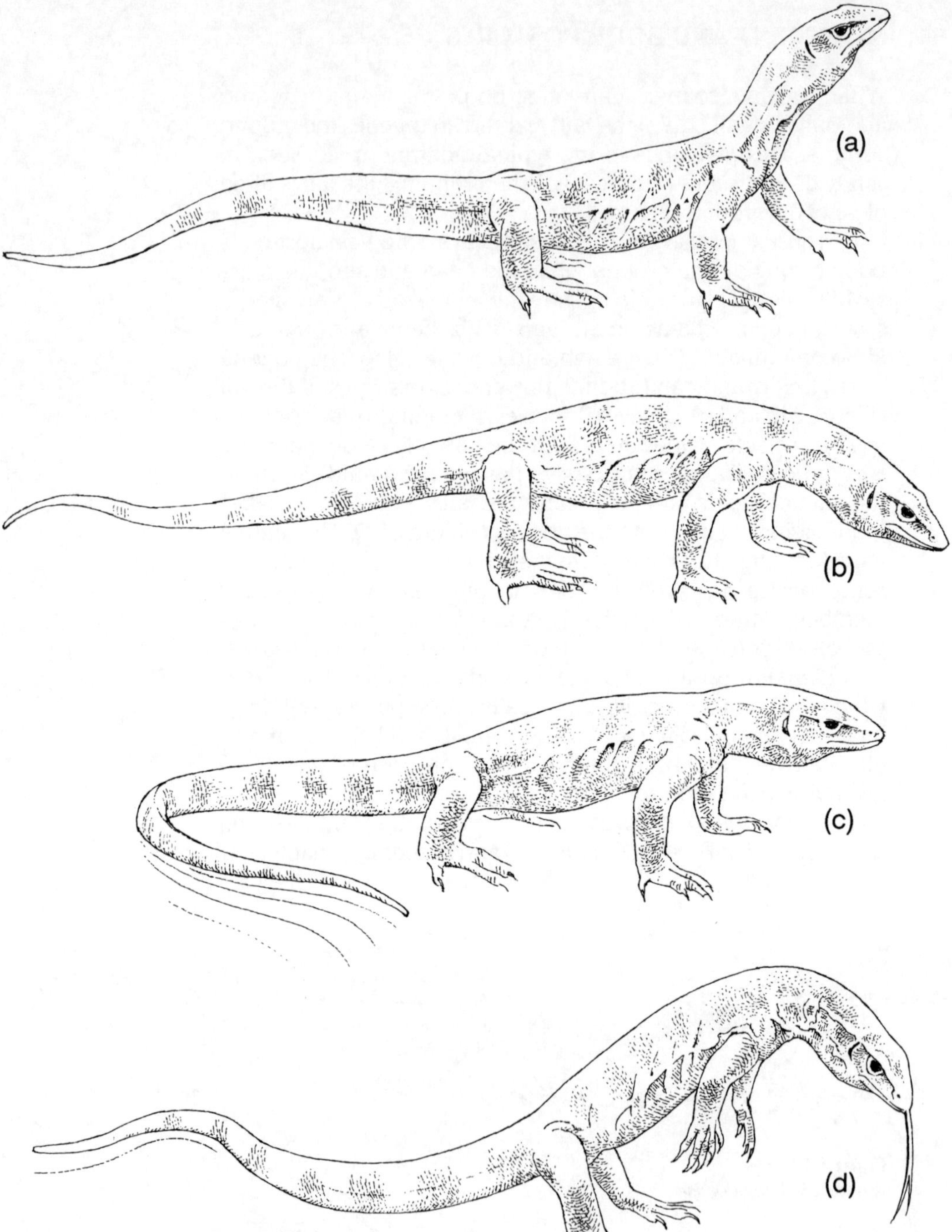

Figure 24
Sequence of postures:
(a) non-aggressive
(b and c) aggressive, with tail swipes
(d) aggressive, upright

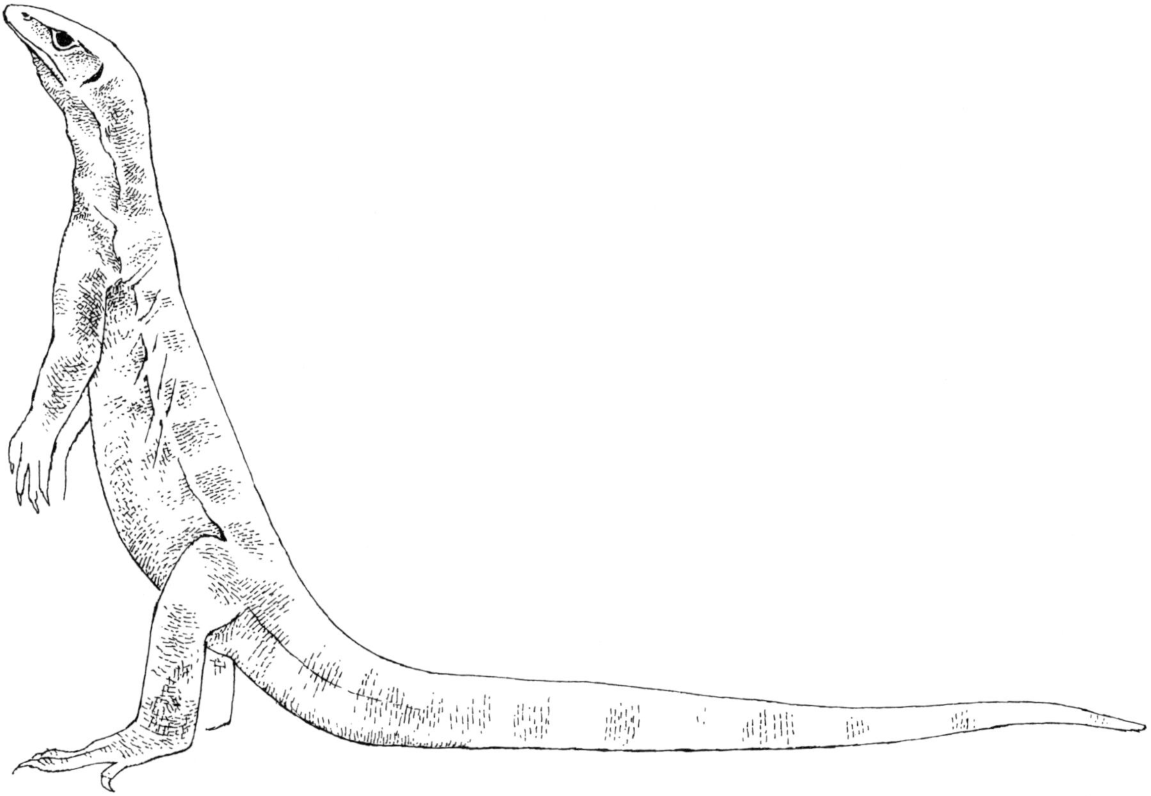

Figure 25
Upright stance of *V. panoptes*

CHAPTER

6

THERMAL BIOLOGY: Regulation of Body Temperature

All varanids are ectotherms. That is, they heat their bodies by basking and absorbing much of their energy directly from the rays of the sun or by conduction from warm surfaces. Goannas of different species maintain their bodies within a temperature range of between 30 and 40°C when they are active (Figure 26), although amphibious species mainly operate at body temperatures near the lower end of that range. Goannas do not regulate their body temperature as precisely as do some other lizards, although most species maintain a body temperature within 2 or 3 degrees of 36°C while active. The body temperatures of animals resting in their burrows are lower than their activity temperatures and can drop to below 10°C in the winter near the northern and southern limits of their distribution.

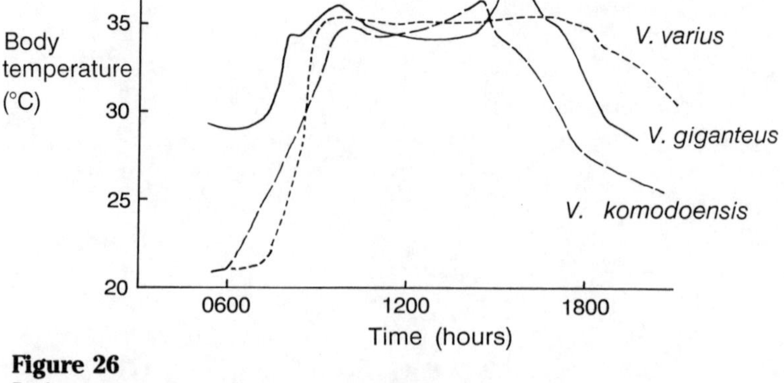

Figure 26
Body temperature profiles of three varanids

BURROWS

Burrows are often used by goannas for protection against overheating on hot days. However, burrows and other forms of shelter, such as holes in trees, hollow logs and brushpiles, also allow body heat to be conserved overnight. The animals can therefore emerge from their shelters in the morning at a higher temperature than their surroundings, and be more active and alert than would otherwise be the case.

Burrows of *Varanus rosenbergi* on Kangaroo Island are relatively shallow, slightly curving, and have an enlarged terminal chamber at the deepest point (Figure 21). Burrows are usually located on flat, open areas during spring and summer. Throughout the year, burrows can also be found on sandy ridges, but during the winter only burrows at these elevated sites are inhabited. The average length of burrows on Kangaroo Island is 105 centimetres and does not vary seasonally, although the depth of the burrows does. Those used during the summer have an average depth below the soil surface of 137 millimetres, while those used during winter are significantly deeper with an average depth of 180 millimetres. The burrows excavated by *Varanus gouldii* during summer in semi-arid habitats are over 400 millimetres below the soil surface. Below soil depths of 150 millimetres, the temperature within the burrow fluctuates very little.

Burrows are used as shelter by many species of goannas (*Varanus eremius, Varanus gouldii, Varanus griseus, Varanus bengalensis* and *Varanus komodoensis*). They vary in shape and are generally relatively short and shallow. Often they are dug by other species and taken over by goannas. Some species also shelter in hollow trees (*Varanus varius, Varanus tristis, Varanus komodoensis*) or in abandoned termite mounds (*Varanus bengalensis*).

The temperatures in the burrows of *Varanus rosenbergi* vary only slightly during the day, reaching a maximum between 4 pm and 10.30 pm, and a minimum temperature between 6 am and 9 am. Burrow temperatures are more stable than air temperatures immediately outside the burrows (Figure 27). In summer, burrows provide cool refuges during the heat of the day, while in winter they remain warmer than the cold night air. At a given time of year, the burrow temperature is approximately the same as the daily mean air temperature outside the burrow. The relative humidity in the burrows remains at or near saturation level even during summer, thereby assisting the goannas to reduce their evaporative water losses.

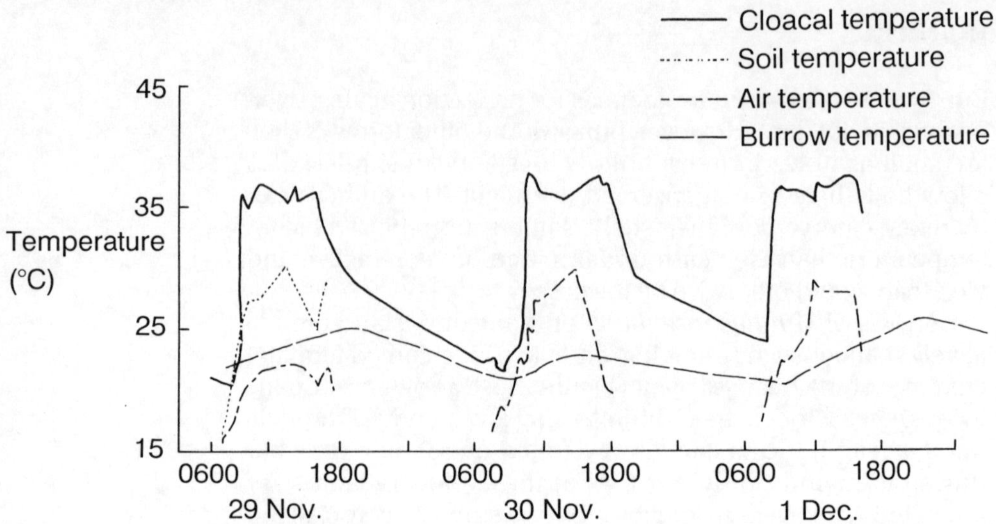

Figure 27
Differences between environmental and body temperatures in *V. rosenbergi*

EMERGENCE TIME

Varanus rosenbergi emerge from their burrows earlier during summer than in winter. Mean time of emergence during summer is 8.50 am (ranging from 7.35 to 10.10 am) while in winter the mean time of emergence is 9.37 am (ranging from 7.45 to 10.45 am). On about 10 percent of winter days goannas on Kangaroo Island do not emerge from their burrows at all. On most winter days they move to the mouth of the burrow in the morning and lie there in apparent anticipation of becoming fully active. The mean body temperature at the time of emergence in summer is 21.8°C (ranging from 15.0° to 24.6°C) while the mean emergence temperature in winter is 12.5°C (ranging from 9.5° to 15.0°C). Temperatures within the burrow are similar to the body temperature of the lizard at the time of emergence.

The precise stimulus for goannas to emerge from their burrows is unknown. It appears that the difference between the temperature inside the burrow and the increasing soil temperature near the burrow entrance may be an important factor. The time they emerge may also vary according to the location or orientation of the burrow. Those animals using burrows near the tops of ridges, or with entrances facing east, generally emerge earlier than goannas using burrows with other locations or orientations. The conditions and times of emergence of other species of varanids are also variable (*Varanus griseus, Varanus gouldii, Varanus giganteus, Varanus komodoensis* and *Varanus varius*).

BASKING

Basking allows the body to gain heat using the radiant energy of the sun. It generally occurs in the morning, after the lizards have emerged from their overnight shelter, but may also occur at intervals throughout the day. Goannas align their bodies in such a way as to maximise heat gain, or elevate themselves above the cold ground by climbing onto logs or rocks to present a larger surface area to the rays of the sun. The heating rates of some species, including *Varanus rosenbergi* (Table 7), vary seasonally. On sunny winter days, the periods of basking are longer and the rates of heating are slower than in the warmer months.

Table 7
Seasonal differences in heating rates and times of basking of *Varanus rosenbergi* on Kangaroo Island

Month	Heating rate (°C/minute)	Basking time (minutes)	Air temperature (°C)
May–June	0.15 (0.04–0.20)	102.5 (85–130)	18.2 (16.3–20.4)
August	0.10 (0.02–0.18)	130.3 (48–265)	15.6 (12.6–19.7)
November–December	0.27 (0.11–0.44)	61.4 (35–95)	22.1 (18.0–26.1)
March	0.28 (0.17–0.49)	60.0 (15–110)	29.5 (23.3–33.7)

All values are means and (range).

It is a common practice of many lizards to protrude only their head from the burrow before emerging completely. It has been suggested that this allows the temperature of the head to be raised, which may increase the level of mental activity of the lizards before they emerge fully. No heating of the body occurs at this time, no doubt because the body has a much greater mass and thermal inertia than the head and there is little transfer of heat from the head to the body.

When a varanid has attained a sufficiently high body temperature by basking, it can move away from the vicinity of its burrow and start hunting, or in the breeding season, searching for a mate.

While active, they travel through patches of sunlight and shade. The density of the vegetation on much of Kangaroo

Island provides abundant shade for *Varanus rosenbergi*. Because the goannas are unable to produce sufficient internal body heat, and have little insulation to help retain it, their body temperatures fluctuate slightly during activity periods. Nevertheless, it is advantageous for ectotherms to tolerate minor fluctuations in body temperature. If they were unable to do so, they would waste time and effort in frequent shuttling between sunny and shady areas to regulate their body temperature more finely. This is especially so in smaller species, because the temperatures of these animals vary more widely, due to the high ratio of surface area to body mass.

ACTIVITY TEMPERATURES

Two methods are used for studying the temperatures of free-ranging lizards. Radio-telemetry (the use of temperature-sensing transmitters) enables a number of temperatures from one animal to be obtained every few minutes while the animal is active in its normal environment. This allows the temperature regulation of an individual to be examined. The other method involves the use of thermometers. This requires the frequent recapture of an individual and alters its behaviour pattern while it is being observed between captures. To overcome this, a number of animals are captured only once each. Consequently, data obtained using thermometers are of less value than transmitter data, as they provide only an average of single values obtained at the time of capture from a number of different individuals.

Under favourable conditions, active *Varanus rosenbergi* on Kangaroo Island maintain a mean body temperature of 35.6°C, ranging from 32.1° to 38.0°C (Figure 27). These temperatures were determined by using radio transmitters on three goannas, releasing them into the wild, and recording their body temperatures while they were active over several days. The mean body temperature of active animals measured with a thermometer shortly after capture during summer was 35.1°C (ranging from 32.4° to 37.6°C).

Studies of temperature regulation of reptiles in the laboratory are generally conducted in artificial temperature gradients. These usually consist of a narrow cage up to several metres long with a heat source at one end. This produces a gradation in air and surface temperatures along the length of the cage. The animal is thus able to select a temperature at any position along the gradient and to regulate its body temperature at its desired level. The thermal environment can be changed to allow particular aspects of the animal's behavioural and physiological reactions to be observed.

The mean body temperature selected by active *Varanus rosenbergi* in temperature gradients was 35.4°C, which is very similar to the temperatures selected by this species in the wild.

Lizards in the wild often become active at lower body temperatures than those selected by captive animals in thermal gradients. Their body temperatures are influenced by the prevailing ambient temperature and degree of solar radiation (insolation). During the colder months, lower air temperatures and less intense solar radiation may prevent them from reaching their normal activity temperatures (Table 8). In summer, goannas avoid becoming overheated by seeking shade, as excessive body temperatures (over about 43°C) will cause death. Free-ranging goannas minimise the range of environmental conditions they encounter by choosing the appropriate time of day to be active.

Table 8
Seasonal differences in activity temperatures of *Varanus rosenbergi* on Kangaroo Island

Month	Number of values	Mean body temperature (°C)	Mean air temperature (°C)
May–June	34	29.3 (18.4–36.8)	15.8 (14.2–18.3)
August	28	24.5 (17.1–37.5)	14.5 (12.4–16.3)
November–December	179	33.4 (21.3–39.2)	22.3 (16.2–33.0)
March	36	32.0 (27.5–36.3)	29.5 (18.5–37.7)

All values are means and (range).

The activity temperatures of *Varanus rosenbergi* are similar to those of most other terrestrial or arboreal varanids. The temperatures of active, free-ranging goannas of five other species (*Varanus giganteus, Varanus varius, Varanus griseus, Varanus gouldii* and *Varanus komodoensis*) have been determined using radio-telemetry (Table 9). Values determined for other terrestrial or arboreal species using single values obtained with a thermometer are shown in Table 10. The body temperatures of active varanids are similar in species that vary greatly in body size.

The mean activity temperatures of three free-living amphibious varanids (*Varanus niloticus, Varanus salvator* and *Varanus mertensi*) are several degrees lower than those of the more terrestrial species (Table 10). These animals are frequently submerged in water that is much cooler than the air and the animals would have to spend excessive periods basking in order to maintain the higher body temperatures commonly found in terrestrial species.

Table 9
Activity temperatures of 5 species of free-living terrestrial and arboreal varanids, as determined by telemetry

Species	Mean temperature (°C)	Temperature range (°C)
V. giganteus	35.8	26.8–39.4
V. gouldii	35.5	27.2–38.1
V. griseus	36.8	27.0–41.6
V. komodoensis	35.1	27.6–41.3
V. varius	35.5	32.8–36.4

Table 10
Activity temperatures of 12 species of terrestrial and arboreal, and 3 species of amphibious, varanids as determined from single values taken by thermometer

Species	Free-living (°C)	Captive (°C)	Total length (m)
TERRESTRIAL OR ARBOREAL			
V. bengalensis	32–37	–	2.0
V. caudolineatus	37.8	–	0.25
V. eremius	37.5	35.9	0.45
V. exanthematicus	36.4, 34.9	36.5	2.0
V. giganteus	36.1	–	2.4
V. gilleni	37.4	37.1	0.7
V. gouldii	37.0	35.3	1.0
V. griseus	38.5	36.4	1.3
V. komodoensis	36–40	36.3	2.8
V. rosenbergi	35.1	35.2	1.5
V. tristis	34.8	35.4	0.8
V. varius	34.7	33.5	2.0
AMPHIBIOUS			
V. mertensi	32.7	32.5	1.3
V. niloticus	32.7	34.8	2.0
V. salvator	27–32	35.6	2.5

HEAD–BODY TEMPERATURE DIFFERENCES

Some researchers believe that lizards regulate their head temperatures more carefully than their body temperatures, in order to prevent overheating of the brain. Temperature gradients occur within the bodies of reptiles and they are maintained by adjustments to blood circulation and respiration. Thermal inertia of tissues and differences in the surface area/volume ratios of different parts of the body are also involved.

Head and body temperatures in several *Varanus rosenbergi* and *Varanus gouldii* in thermal gradients were measured simultaneously. The temperature inside the skull of the goannas was generally slightly lower than the cloacal temperature before the animals began basking. After the lizards emerged from artificial burrows into the gradient and started to bask beneath heat lamps, their head temperatures increased more rapidly than their body temperatures. After 20–40 minutes of basking, body temperature rose above cranial temperature and remained above it for the remainder of the animals' activity period. When heat was no longer applied, the heads and bodies of the lizards cooled at similar rates (Figure 28). When measured over a period of weeks, the average internal head and body temperatures during activity were similar (Figure 29), as were the maximum temperatures. If only the body of a goanna is heated, both head and body temperatures rise, whereas if only the head is heated its temperature rises while the body temperature does not. This may explain why goannas protrude their heads from their burrows for a short time before they emerge completely.

A study using transmitters on free-ranging *Varanus varius* showed that brain temperatures were higher than body temperatures by as much as 6°C before emergence, or during the initial period of basking. However, once the animal became active, brain and body temperatures were similar.

The differences in the mass/surface area ratios between the heads and bodies of varanids and the blood flow between them are responsible for the differences in the rates of heating of the head and body. Evaporative cooling of the moist surfaces of the mouth and the eyes also affects the temperature of the head.

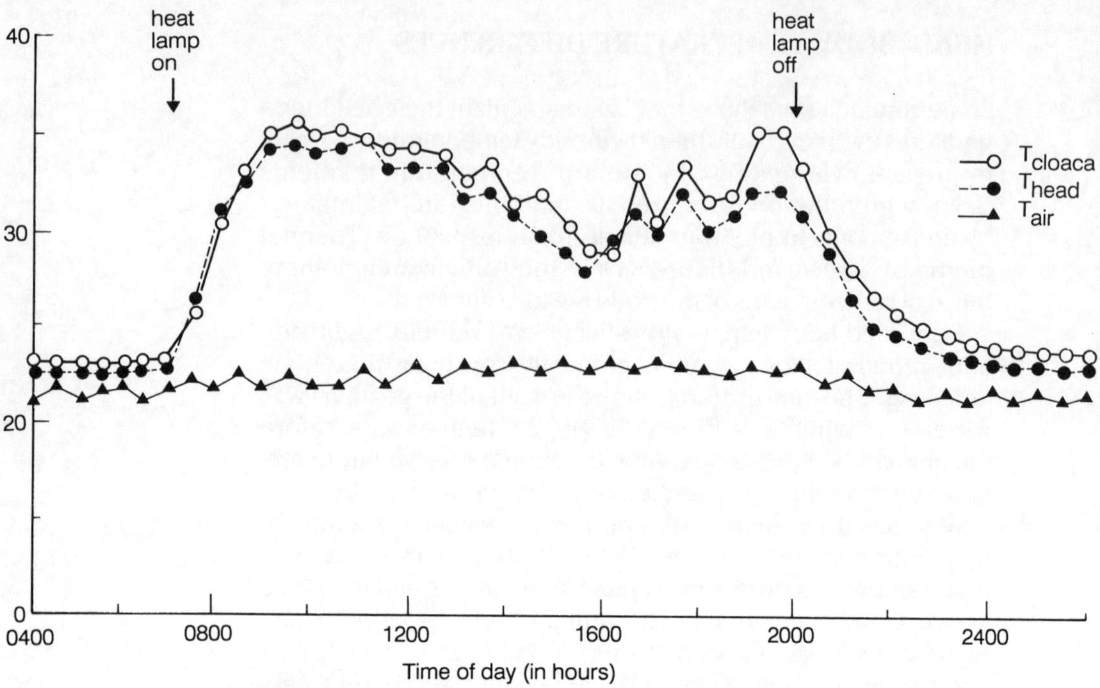

Figure 28
Head and body temperatures of *V. rosenbergi* while basking and active during one day

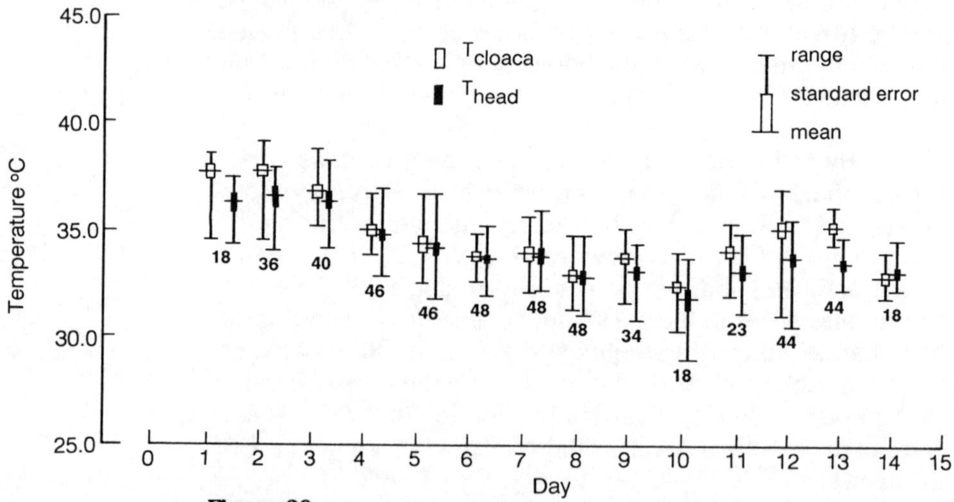

Figure 29
Average head and body temperature differences of active *V. rosenbergi* over 2 weeks in a thermal gradient

NASAL TUBES

The anterior nasal tubes have a large surface area and this is important in heat exchange and water conservation. Inhaled air is warmed and humidified as it passes along the nasal tubes. As exhaled air passes back along the nasal passageways, which are now cool, heat is transferred from the air back to the tissues. Water vapour is condensed onto the surface of the nasal tubes before it leaves them. Thus water loss is less than it would be if expired air remained at deep body temperature. In some species, including *Varanus rosenbergi*, the nostril is near the tip of the snout. This means the nasal tubes are as long as possible, which aids in warming the inhaled air. In some other species the nares are immediately in front of the eyes, and the nasal tubes are short.

GULAR FLUTTERING

Gular fluttering, which is often erroneously referred to as panting, is used by goannas and by many other types of lizards to cool themselves when they become overheated. The mouth is held open and the gular region (Figure 1) of the throat is fluttered rapidly. This has the effect of increasing the evaporation rate from the mouth and throat area and therefore reducing temperature. However, it is not often employed as a means of cooling because water losses may increase by 4–6 times when it occurs (see Water Use, Chapter 8 and Figure 32). Nevertheless, it is a means of reducing dangerously high head temperatures in an emergency. Lizards normally avoid these problems by retreating from stressful temperature conditions to areas of shade, or into burrows.

When *Varanus rosenbergi* were placed in temperature-controlled cabinets with rising temperatures, they began gular fluttering when body temperature rose above head temperature (about 38°C). This is similar to the temperatures at which *Varanus gouldii*, *Varanus griseus* and *Varanus komodoensis* begin gular fluttering. Gular fluttering is not always continuous, but usually occurs at a rate of 60–90 per minute and reaches up to 120–130 per minute. Its onset was occasionally preceded by a period of struggling by the lizard. Head temperatures either stabilised or dropped slightly when gular fluttering began, although body temperatures continued to rise. Head temperatures rose during the brief periods when fluttering stopped. Stable head temperatures that were as much as 2.8°C below body temperature were recorded, and these were maintained for periods of up 75 minutes. Head temperatures were maintained below ambient air temperature, but body temperatures were not lower than ambient temperature during gular fluttering. No changes occurred in the depth of respiratory movements, which indicates that the function of gular fluttering is not respiratory but serves to control the temperature of the brain.

REFLECTIVITY OF THE SKIN

Skin colour is important in thermoregulation as well as in breeding activity and camouflage. Some lizards can change their skin colour under different environmental conditions to increase or decrease the rates at which they take up or lose heat. Varanids, however, are unable to change colour. Differences in skin colouration and reflectivity between species are related to the habitats they use. *Varanus rosenbergi* inhabits densely vegetated areas, and is a darkly coloured (melanistic) species.

Nearly all of the light energy which reaches the earth's surface is in the visible or infra-red parts of the spectrum. Some of it is absorbed by lizards when they bask. The amount of light reflected by skin from *Varanus rosenbergi* (from Kangaroo Island) and the light coloured *Varanus gouldii* (from arid and semi-arid regions) has been compared. The greatest differ-

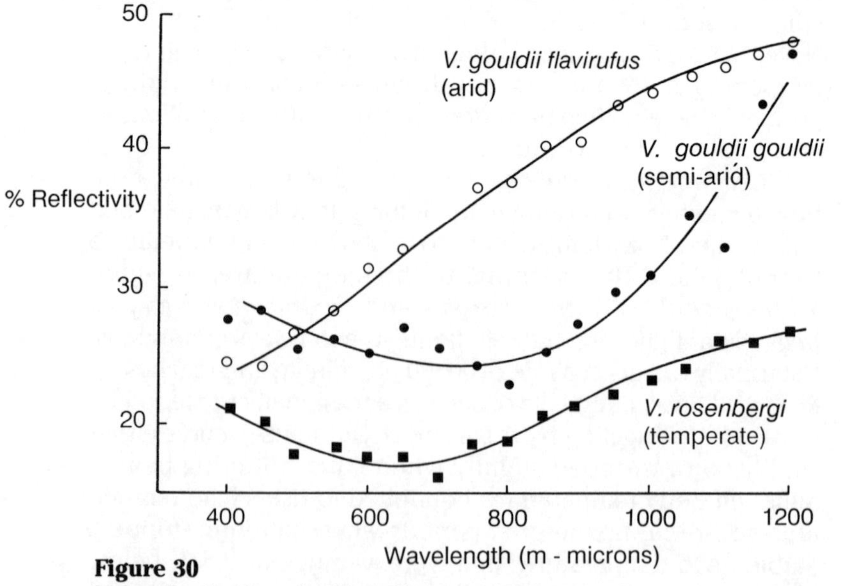

Figure 30
Light reflection from the skins of *V. rosenbergi* and *V. gouldii*

ences in reflectivity found in goannas from the three habitats were in the near infra-red portion of the spectrum (Figure 30). Skin from *Varanus rosenbergi* reflected less light than the others in all parts of the spectrum.

Heavy, moist soils and those covered with heavy plant growth reflect the least solar energy. Bare sandy soils reflect the most. As there is usually abundant sunlight in deserts, temperatures there are high. Lizards that live in such environments can easily become overheated, but they can remain active for longer periods if their skin reflects large amounts of this energy. The temperatures on Kangaroo Island are lower than those in the arid and semi-arid regions of Australia, and the vegetation is also more dense than in the deserts. During much of the year there is not enough sunny weather to allow goannas to easily reach their activity temperatures when they bask. The dark skins of goannas living on Kangaroo Island absorb more solar energy, which means the animals can heat up more rapidly and be active longer in the colder, wetter months.

CHAPTER 7

RESPIRATION

Living cells require a constant supply of oxygen to enable them to respire and maintain normal cellular functions. Oxygen must be transported to the cells and the carbon dioxide produced in respiration must be carried away and expelled from the body. This process of gaseous exchange is carried out by the lungs and circulatory systems of animals.

In varanids, these systems differ in a number of ways from those of other reptiles. Reptiles, unlike mammals, do not possess a diaphragm (a sheet of muscle which separates the organs of the chest from the organs of the abdomen). In most reptiles the heart and lungs are not enclosed in a separate chest cavity but are located in the abdominal cavity with other organs (Figure 13). However, varanids and their relatives, *Lanthanotus* and *Heloderma*, have a septum (a dividing membrane) which separates the abdominal and chest (pleural) cavities. Except for crocodiles, reptiles do not seem to have specialised breathing muscles. They breathe in by drawing the ribs upwards and forwards by means of the intercostal muscles (situated between the ribs), increasing the size of the pleural cavity. When those muscles relax, the weight of the body and the elasticity of the muscles and other tissues forces air out of the lungs.

LUNGS

Reptile lungs have a volume approximately six times larger than the lungs of similar sized mammals, and are also larger than the combined lung–air sac systems of birds. However, the lungs of goannas are less elaborate than those of birds and mammals. Varanid lungs are larger than those of most other reptiles, thus enabling them to extract more oxygen from the air they breathe.

Reptilian lungs have to function over a wide range of oxygen consumption rates, due to the great differences in the animals' activity levels. Even so, these rates are much lower than those of birds and mammals which generally have much higher energy requirements.

Each lung has a cartilage reinforced intra-pulmonary duct (the bronchus) which connects with a number of small chambers. This is called a multicameral lung (Figure 31). It provides an enlarged surface for gas exchange. This in turn provides an advantageous starting point for the evolution of a high performance lung containing a large surface area which can be ventilated at a low energy cost. Exchange of oxygen and carbon dioxide between the inspired air and the air in the lungs is thus greatly enhanced.

Varanid lungs are attached directly to the body wall where it underlies the rib cage. This prevents them from collapsing completely because of over-stretching during normal breathing. They have a strong compliance with body movements and thus can be efficiently ventilated by means of costal breathing alone. This results from the normal movement of the ribs altering the volume, and thus the pressure, in the lungs. This and their multicameral structure produces a more efficient system of air circulation than that of other lizards. The lungs of most lizards lie free in the body cavity and must be actively deflated, thereby increasing the energetic costs of breathing. Varanids can extract oxygen from inspired air at rates that are similar to those of mammals at rest. Varanids also have a high aerobic capacity (capacity for aerobically supported activity which relies on oxygen) and breathing requires little extra expenditure of energy.

When active, most reptiles rely heavily on anaerobic respiration, which does not require oxygen. Instead, energy is released by the partial breakdown of stored carbohydrates to form lactic acid. Unlike aerobic respiration, anaerobic respiration has the advantage of being equally efficient over a wide range of body temperatures. Anaerobic respiration is therefore useful for reptiles which must be active before they have warmed up to their normal activity temperatures. However, it has the disadvantages of releasing only a small part of the energy available in the carbohydrate energy source, and of generating large quantities of lactic acid, which accumulates in the muscles and prevents sustained activity. Lactic acid is slowly removed from muscles by the blood and is later broken down aerobically.

However, active varanids mainly rely on aerobic respiration and therefore do not generally build up high lactic acid levels in their tissues. They are therefore able to sustain high rates of activity over long periods of time.

The lungs of varanids receive a regular supply of blood, unlike those of other lizards, where blood is shunted away from the lungs during long periods of breath-holding. The breathing rate of varanids is generally regular and low, but it can be variable. Varanids can voluntarily hold their breath for long periods. For example, *Varanus exanthematicus* in the laboratory have been observed to hold their breath for up to 8 minutes. Even longer periods of breath-holding have been recorded for amphibious varanids.

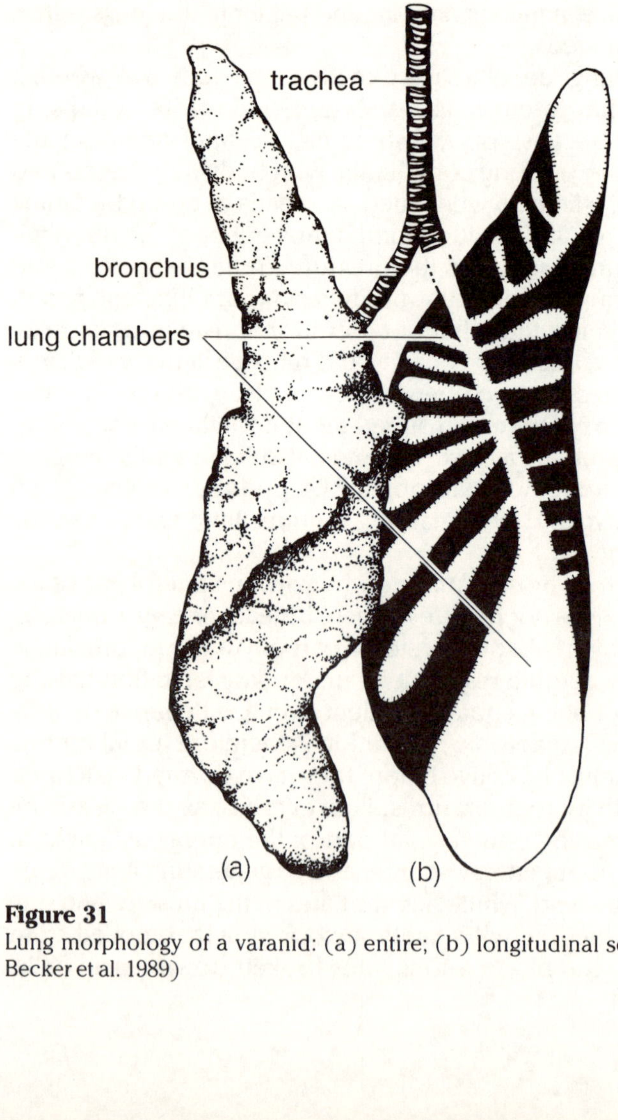

Figure 31
Lung morphology of a varanid: (a) entire; (b) longitudinal section (after Becker et al. 1989)

HEART

The heart in varanids is located further back in the body than it is in most lizards. It lies just behind the posterior part of the sternum (breastbone), embedded between the lobes of the liver. The anterior region of the heart is attached to a thick pericardium (a membranous sac which encloses the heart).

The varanid heart is not a typical lizard heart. The arrangement of the chambers and valves results in only a small amount of mixing taking place between the two bloodstreams. About 30 percent of the blood intended for the lungs goes instead to the body and approximately 10 percent of the oxygenated blood is returned to the lungs. This is a much lower level of mixing than found in most lizards, however the structure of the heart appears to make some mixing of the bloodstreams unavoidable. The arterial (oxygenated) blood in varanids carries higher levels of oxygen than it does in most other lizards.

The relatively long necks in varanids may impose additional circulatory demands similar to those of snakes. Both types of animals require a supply of blood to the body at high pressure. However, the blood supply to the lungs is at a much lower pressure. These pressure differences are similar to those found in the circulatory systems of mammals. They are achieved by means of a muscular ridge in the ventricle which presses tightly against the heart wall at the appropriate phase of the heartbeat. This effectively creates a four-chambered heart and separates the two systems.

BLOOD PHYSIOLOGY

As the body temperatures of reptiles rise during activity, species that rely heavily on anaerobic respiration release large amounts of lactic acid into their blood. Increased acid levels raise the acidity of the blood unless an efficient buffering system is present to prevent it. This system must function effectively over a range of body temperatures. A high acidity of the blood reduces its oxygen-carrying capacity, which in turn increases the reliance of a lizard on anaerobic respiration. This in turn leads to the formation of yet more lactic acid which further decreases the oxygen-transporting ability of the blood.

The high level of aerobic respiration of varanids is partly due to the different way in which the buffers in their blood function. Varanids closely regulate the acidity of their blood as their body temperature increases and they become more active. The lungs of varanids are highly efficient in ridding the body of excessive carbon dioxide, which helps maintain the function of buffers to prevent an increase in blood acidity. Their lactic acid production is also maintained at low levels. Goannas are thus able to maintain their activity for long periods.

Varanids also have higher levels of myoglobin than other lizards. Myoglobin is a protein in the muscle which binds with oxygen transported there by the blood. These levels are similar to those in mammals, and improve the rate of oxygen transfer from the blood to the muscles.

The blood supply to the lungs also increases when body temperature rises. Thus varanids are very mammal-like in many features of their cardio-pulmonary physiology. This includes the need to maintain high levels of oxygen intake when they have high body temperatures but are at rest.

The respiratory properties of the blood of the amphibious species of varanids show some adaptations to extended submergence. The buffering capacity of their blood allows them to be more tolerant of lactic acid accumulation than are other species of varanids. High lactic acid levels will initiate breathing and thus restrict the length of diving periods. Dives of as long as 1 hour have been observed in some amphibious species, as a result of their higher tolerance to lactic acid. During diving, the body temperatures of amphibious species drop rapidly. To overcome this, the blood is able to release oxygen to the tissues at lower body temperatures than occurs in terrestrial species.

The advanced structure and performance of the heart and circulatory system, in conjunction with the complex structure of the lungs, allows varanids to respire efficiently for long periods without them becoming exhausted. Thus they are able to adopt an active, wide-ranging foraging strategy that is suited to their feeding habits.

CHAPTER 8

WATER USE

Six species of goannas live near water, and they can be regarded as essentially amphibious (*Varanus indicus, Varanus karlschmidti, Varanus mertensi, Varanus niloticus, Varanus salvator* and *Varanus semiremex*). These species are usually found in freshwater habitats, but can extend into brackish waters, or even sea water. This is especially true of those monitors which inhabit mangroves (*Varanus indicus, Varanus salvator* and *Varanus semiremex*). When in freshwater, these species have unlimited quantities of drinking water and therefore have no problems with maintaining water balance. Another species (*Varanus mitchelli*) also uses river pools, but is primarily arboreal, and frequents trees on the edges of small creeks in northern Australia, occasionally taking to the water to escape capture.

Wild animals will drink water if it is available. This is especially so for herbivores living in hot or arid environments. However, carnivorous animals obtain a large amount of water from their food, as about 85 percent of the body of prey animals is available as water. Because of this, carnivores need less drinking water than most herbivores. Goannas have been seen to drink in the wild, although those that live in deserts rarely have the chance to do so. Desert species are frequently subjected to shortages of free water. How do terrestrial and desert varanids cope with these water shortages and with the high levels of salts that they may ingest?

WATER LOSSES

Terrestrial animals lose water to their surroundings in a number of ways. These include evaporation from the lungs during breathing, evaporation from the skin, and loss of water in the urine and faeces. Reptiles, particularly those living in arid environments, show a number of adaptations for restricting their water losses.

Evaporative water loss

The amount of water that a goanna loses by evaporation depends on a number of factors. The most important ones are air temperature and humidity, body temperature, and the level and type of activity the animal is engaged in. Laboratory studies have measured the evaporative water loss of *Varanus rosenbergi* in a dry atmosphere. These show that as the temperature rises from 26° to 38°C the loss of water from the skin and lungs gradually increases by about three times (Figure 32). This is because the metabolic rate and respiration rate both increase as the body temperature rises. The air also has more 'drying power' (saturation deficit) at higher temperatures.

When their body temperatures rise above approximately 38°C, goannas begin to flutter the gular pouch under their

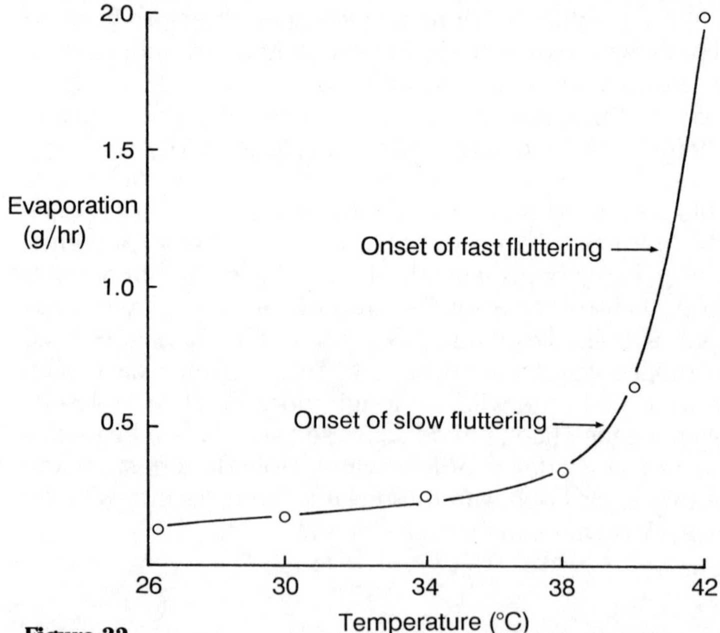

Figure 32
Effect of gular fluttering on evaporative water loss in *V. rosenbergi*

throat (Figure 32). This causes a dramatic increase in the rate of evaporation from the lungs and mouth cavity. Initially, gular fluttering is shallow, at a rate of 60–90 movements per minute, with the mouth held open only slightly. Even so, total evaporation almost doubles. As body temperatures increase, gular fluttering rates increase to more than 120 per minute, and there is a further four-fold increase in total evaporative water loss (Figure 32).

Obviously goannas could not sustain such high rates of water loss for long periods of time, as they would rapidly dehydrate. Gular fluttering is a useful short-term method for losing heat from the head and brain, but it is rarely used under natural conditions. In the field, goannas would seek out cooler microclimates, such as burrows or deep shade, to avoid or reduce body heat loads (see Thermal Biology, Chapter 6).

The skins of terrestrial reptiles are dry to the touch as they do not contain sweat glands. The skin is thickened with the protein keratin, which, together with lipids, helps to minimise evaporation from the skin (cutaneous water loss). Their rates of water loss through the skin are extremely low compared with those of other terrestrial vertebrates (Table 11).

Evaporation from the skin increases as the temperature rises. *Varanus rosenbergi* loses 0.12 milligrams of water per hour through each square centimetre of its skin at 30°C. The evaporation rate doubles when its temperature rises to 38°C, due

Table 11
Rates of water loss from the skin of some lizards and other vertebrates at 30°C

Species	Water loss ($mg\ H_2O/cm^2/hr$)
LIZARDS	
Varanus rosenbergi	0.12
Varanus gouldii gouldii	0.11
Varanus gouldii flavirufus	0.10
Varanus storri	0.12
Varanus gilleni	0.06
Sphenomorphus	0.25
Gehyra	0.21
Amphibolurus	0.10
Anolis	0.19
OTHER VERTEBRATES	
Painted quail	0.95
Sheep	1.24 (2.26*)
Man	1.48 (11.6*)
Potoroo	1.7

* While sweating

mainly to a higher saturation deficit of the air at higher temperatures (warmer air can hold more water vapour). It is also possible that increased blood circulation to the skin may increase the rate of evaporation.

Large animals have a relatively smaller surface area (square centimetre of skin per gram of body weight), than small animals. The relationship between surface area and body weight in *Varanus rosenbergi* is shown in Figure 33.

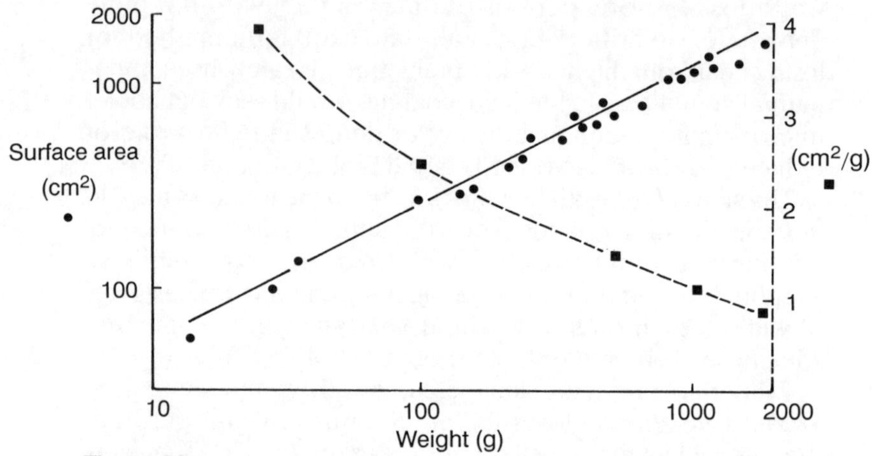

Figure 33
Relationship between skin surface area and body weight in *V. rosenbergi*

The surface area to weight ratio of a hatchling goanna is about five times that of an adult. Because of this, a goanna hatchling would lose far more water through the skin, relative to its body size, than would an adult of the same species. Juveniles are much more vulnerable to desiccation in hot, dry conditions, and they must behave differently from adults if they are to survive in these circumstances.

About 70 percent of all evaporative water loss by inactive goannas is through the skin. Another area for evaporative water loss is the surface of the eyes, which are effectively free water surfaces kept wet by secretions of the tear glands (lachrymal glands). Laboratory experiments show that at 30°C in dry air, *Varanus rosenbergi* loses 67 milligrams of water each hour by evaporation from each square centimetre of eye surface. The eyes of young animals of most species are relatively larger than those of adults, and this is certainly true of goannas. Because of this, relatively more water is lost from the eyes of juveniles than from those of adults. The total eye surface of an adult is only four times that of a juvenile although it weighs 100 times more. Thus the water loss per gram of body weight from the eyes of a juvenile is 25 times that of an adult.

The relative importance of the different avenues of evaporative water loss in goannas of different sizes is shown in Table 12. There is comparatively little water loss from the eyes in adults. However, it is interesting to note that they keep their eyes closed while basking, presumably to reduce water losses from this source. In some reptiles, such as snakes and some geckos, the eyes are covered by transparent scales known as spectacles. These reduce evaporation from the eyes to levels similar to that from the skin. Few desert-adapted reptiles have these 'spectacles', which suggests that 'wet' eye surfaces provide better vision than 'dry' spectacles. The water loss associated with moist eyes can be regarded as an acceptable cost to be paid for having improved vision.

The major sources of water loss in resting goannas are the skin and, in juveniles, the eyes. The respiratory water losses of inactive goannas are amongst the lowest recorded for reptiles. However, once a goanna becomes active within its natural environment, water loss from the lungs increases significantly and becomes the major source of water loss, especially when gular fluttering is used for thermoregulation.

Table 12
Effect of size on evaporative water losses of *Varanus rosenbergi* at 30°C

Weight (g)	Surface area		Evaporation (mg / hr)			Total evaporation
	Skin (cm^2)	Eyes (mm^2)	Skin + lungs	Eyes	Total (mg H_2O/g/hr)	(mg H_2O/g/hr)
15	74	22	8	15	23	1.5
150	341	29	50	20	70	0.5
1500	1611	44	300	29	329	0.2

Urinary excretion

Reptiles produce a number of different nitrogenous waste products that result from the breakdown of proteins. These include ammonia and ammonium salts, urea, uric acid and urate salts. Aquatic species, such as turtles and crocodiles, usually have unlimited access to water to replace that lost during excretion. In these species, the soluble waste products, such as ammonia and urea, predominate. In contrast, terrestrial species generally need to conserve water and cannot afford to excrete waste products that need to be voided with large amounts of water. Apart from one exception, terrestrial reptiles cannot produce a urine more concentrated than the blood. Because of this, the major excretory products in these animals are uric acid and urate salts. These waste products are highly insoluble and require very little water for their excretion.

Birds and mammals are able to produce urine that is much more concentrated than the blood. This is possible because the kidney tubules are folded back on each other (Henle's loops) and there is a concentration gradient across the kidney so that the inner regions have a higher osmotic pressure than the outer regions (Figure 34).

The kidneys of reptiles are simple structures compared with those of birds and mammals (Figure 34). There are no loops of Henle and no osmotic gradient across the kidneys of reptiles. No reptilian kidneys are able to produce urine that is more concentrated than blood. One species of agamid lizard (*Amphibolurus maculosus*), living on some salt-lakes in South Australia, does produce a urine that is more concentrated than blood. However, the kidneys are not responsible for this. Instead, the urine is concentrated by special modifications to the cloaca and hindgut. In reptiles the renal artery enters the

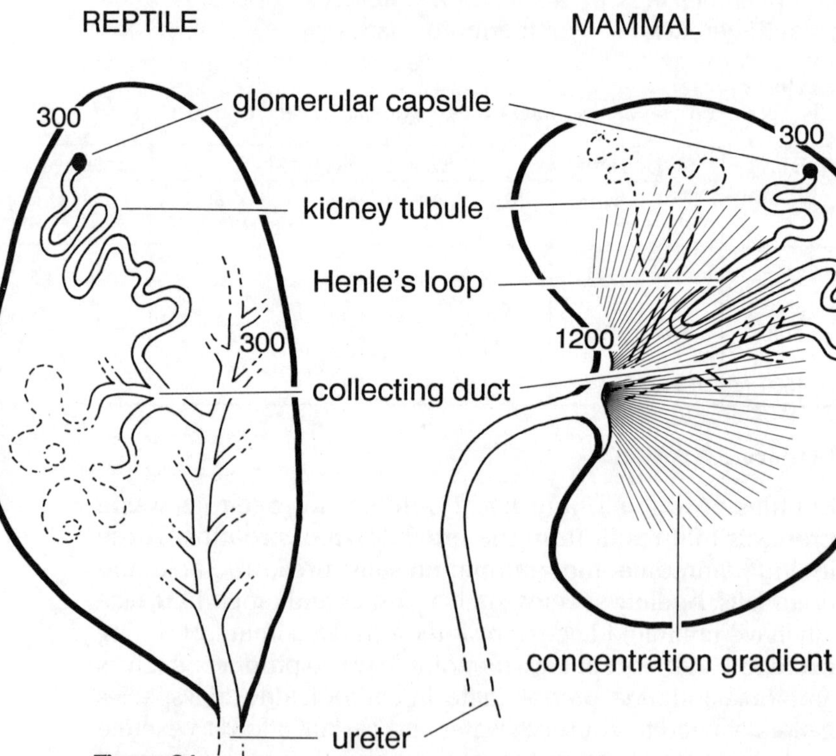

Figure 34
Comparative morphology of a reptile and mammal kidney; numbers indicate approximate osmotic concentrations in milliosmols

kidney and subdivides into arterioles that enter the glomerular capsules (or glomeruli). Here, blood pressure forces water, salts and small organic molecules from the bloodstream across to the other side of the capsule membrane. This process is known as glomerular filtration, and the liquid that passes into the kidney tubules is called glomerular filtrate.

The filtrate passes down the kidney tubules and, as it does so, water and sodium are resorbed across the walls to re-enter the bloodstream. At the same time, nitrogenous wastes in the form of soluble urates and some uric acid are secreted into the tubules, along with some potassium. By the time the filtrate reaches the ureters it is regarded as urine. Normally, the kidneys of *Varanus rosenbergi* produce about 7 millilitres of filtrate and 3 millilitres of urine each hour when the goanna's body temperature is 30°C. However, the amount of filtrate and urine produced can vary, depending on the animal's state of water balance and the amount of salt in the body.

Goannas that are dehydrated or burdened with high salt loads reduce the number of kidney tubules in use, reducing the amount of filtrate produced by the kidneys. The animals also resorb more filtrate in the tubules and produce smaller amounts of urine (Figure 35). However, if a goanna takes in excessive water, and is water-loaded, the number of working tubules and the volume of filtrate produced are increased. Also, less filtrate is resorbed across the tubule walls. This means that larger quantities of urine are produced and the excess water is excreted. Thus the kidneys are vitally important in maintaining the water balance of the goanna.

In reptiles, changes in the number of working kidney tubules, and the extent to which the tubules resorb filtrate, are regulated by an antidiuretic hormone, arginine vasotocin (AVT).

In laboratory experiments, the levels of AVT in the bloodstreams of water-loaded *Varanus gouldii* are low. Wild *Varanus gouldii* in summer have more than double the AVT concentration of water-loaded animals, and levels in salt-loaded animals are double that again. There is a clear correlation between circulating levels of antidiuretic hormone and the state of hydration of the animal. When the level of AVT is low, it is likely that all of the glomeruli are working, with only a small amount of filtrate being absorbed into the tubules. This increases urine production (diuresis). At 'normal' AVT levels, the number of functioning glomeruli in the kidney is reduced, and less filtrate and urine are produced. If AVT concentrations are high, as they are in a dehydrated or salt-loaded animal, there is an increase in tubular resorption, resulting in the production of smaller volumes of more concentrated urine.

If high levels of AVT are injected into goannas, even those which are water-loaded, the glomeruli shut down completely, and no urine is produced for some time. When the kidneys commence functioning again, both filtration rate and urine flow are greatly reduced (Figure 35). Thus, by means of the action of AVT, stimulated by environmental influences, goannas are able to regulate the volumes and concentrations of the urine they produce.

In terrestrial vertebrates the urine leaving the ureters is not immediately expelled from the body. Instead, it is diverted into storage organs such as urinary bladders or cloacas. In reptiles, these storage organs are able to extensively modify the composition of the urine before it is finally excreted.

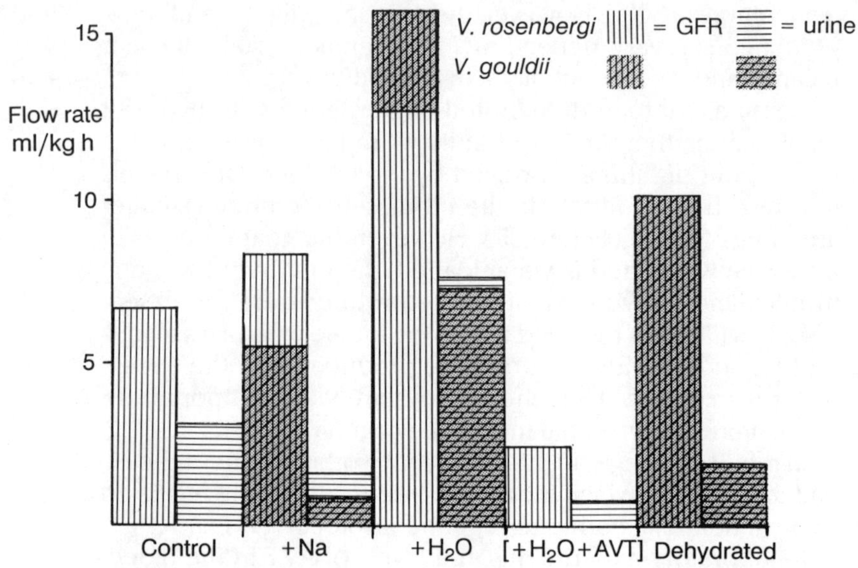

Figure 35
Kidney function in *V. rosenbergi* and *V. gouldii*

Cloacal function

The cloaca plays an important role in resorbing water from urine. In most lizards, the cloaca is divided into three main regions. These are the coprodaeum, which is immediately posterior to the hindgut, the urodaeum and the proctodaeum, which opens to the outside (Figure 36). The paired ureters open into the cloaca via papillae positioned on the dorsal wall of the urodaeum. Urine emerges from the papillae, and passes forward into the coprodaeum, which plays the major role in water resorption.

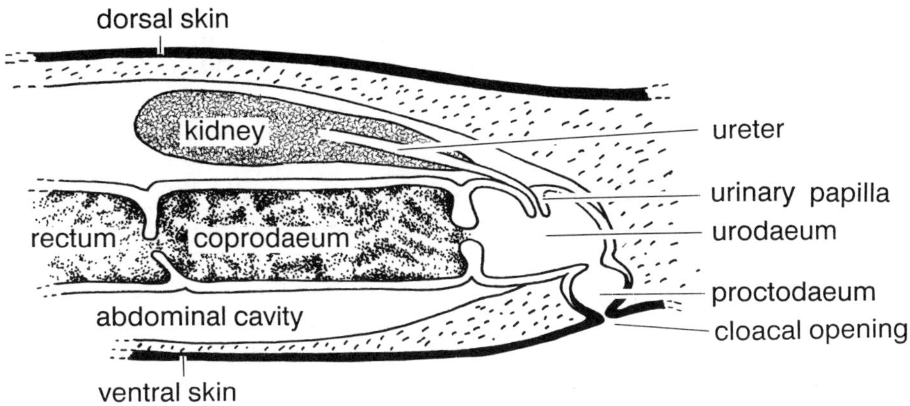

Figure 36
Cloacal structure of *V. rosenbergi* (longitudinal/vertical section)

The coprodaeal chamber is the largest of the cloacal regions. Its lining is highly folded, substantially increasing the surface area of the chamber. Under a microscope, the surface area is seen to be further increased by small projections, known as villi, which are similar to those found lining the absorptive surfaces of the intestines (Figure 37). At even higher magnification, even smaller projections, called micro-villi, are found on some of the surface (epithelial) cells. These give a 'hairy' appearance to the surface of these cells.

The larger the surface area, the greater the amount of resorption that can take place. The folds, villi and micro-villi greatly increase the surface area available for the resorption of fluids from the urine. The cloacal wall also has a rich blood supply which ensures the rapid transport of these fluids out of the cloaca.

Numerous goblet cells in the wall of the cloaca acidify the urine when it enters the coprodaeum. This causes soluble urate salts to become insoluble and precipitate, releasing more water to be resorbed.

The cloaca of *Varanus rosenbergi* resorbs about eight millilitres of water every hour when the animal is normally hydrated. When a goanna is injected with AVT, it is able to resorb about 21 millilitres of water per hour. Urine in the ureters of goannas is more dilute than is the plasma of the blood, even in salt-loaded or dehydrated animals. Because of this, water in the urine can be resorbed across the cloacal wall into the bloodstream by osmosis. This process continues until the urine reaches the same osmotic concentration as the blood. The cloaca can also actively transport sodium out of the urine. This releases even more water for resorption. Eventually, as a result of the precipitation of urate salts and resorptive processes, the solid wastes form a large pellet of uric acid and insoluble urates within the coprodaeal chamber.

By the time the urinary pellet is voided from the cloaca, the water content of the pellet is only 48 percent in normally hydrated goannas. This represents a water loss of only 2.7 millilitres for every gram of nitrogen excreted. This is a much smaller excretory water loss than that achieved by mammals. The most concentrated urine produced by any mammal is that of a desert rodent, *Notomys alexis*, which has an excretory loss of seven millilitres of water for every gram of waste nitrogen. It is likely that dehydrated goannas produce even drier pellets, further reducing their excretory water losses.

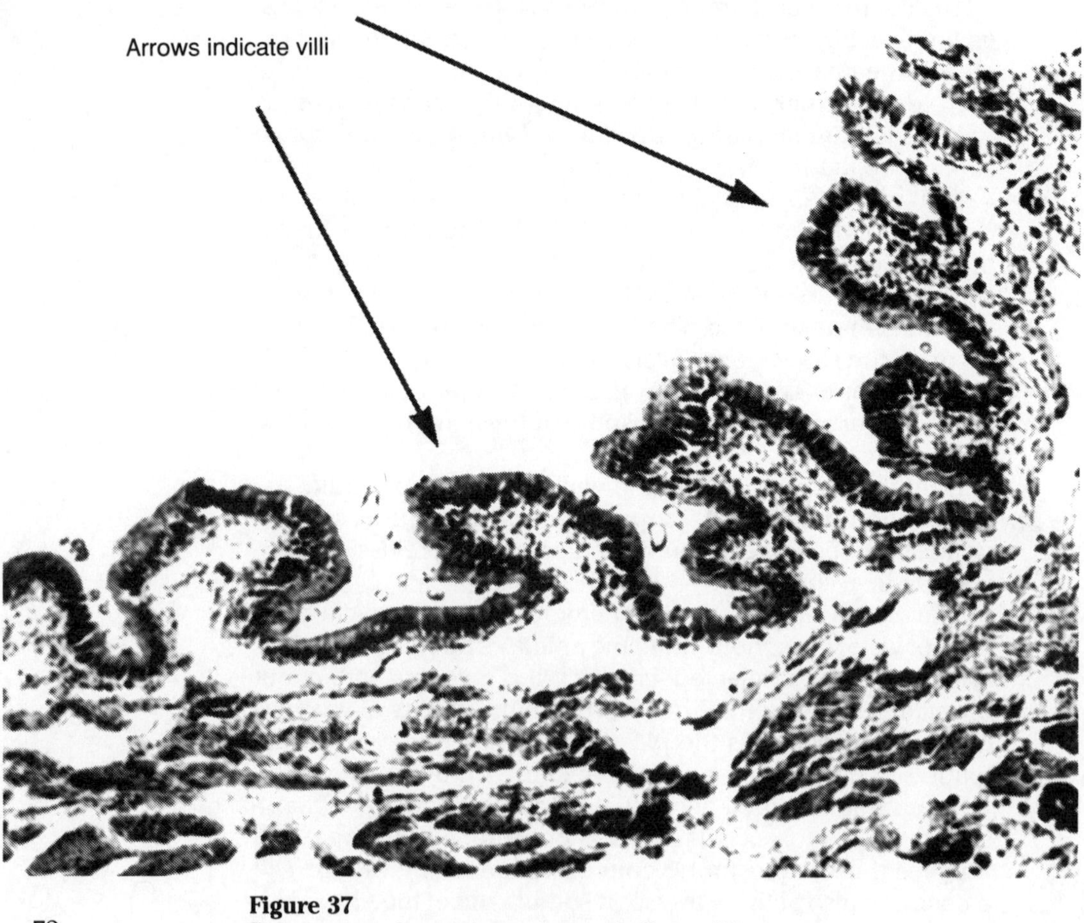

Figure 37
Fine structure of coprodaeum (arrows indicate villi)

The faeces are formed in the posterior region of the hind-gut, immediately anterior to the coprodaeum. This region of the gut has a similar wall structure to that of the cloaca, and resorbs water from the faecal matter. The water content of voided faeces from normally hydrated goannas is 77 percent. This is much higher than that of the urinary pellets. However, drier faeces are produced when animals become dehydrated.

Like other terrestrial reptiles, goannas have an efficient, flexible excretory system that is well adapted to water conservation. The kidneys produce a dilute urine that results from the tubular resorption of sodium and the limited resorption of water. The kidney tubules secrete soluble urates, which prevent the tubules from becoming blocked by insoluble urate deposits. Once the urine has reached the cloaca, the urates are precipitated to form a pellet, allowing the resorption of further sodium and water. In the unlikely event of a goanna becoming water-loaded, the circulating levels of anti-diuretic hormone (AVT) decline and greater numbers of kidney tubules increase the rate of urine production. AVT levels increase as the goannas become dehydrated, resulting in increased rates of water movement across the kidney tubule and cloacal walls, and increased sodium transport in the cloaca.

Since the resorption of urinary water in both the kidney and the cloaca is dependent upon the active transport of sodium out of the urine, how do goannas, especially those species that enter brackish or even sea water, prevent the build-up of sodium within their bodies?

SALT-SECRETING GLANDS

Sodium build-up is prevented mainly by salt-secreting glands, which are found in many fish, birds and reptiles, especially in marine species. They are also present in many terrestrial lizards including members of the Varanidae such as *Varanus rosenbergi, Varanus semiremex, Varanus flavescens, Varanus griseus* and *Varanus salvator*. The salt glands of lizards are in the nasal capsules, those of turtles are lachrymal (behind the eyes), while sea-snakes have sub-lingual and pre-maxillary salt glands in the mouth. It has recently been discovered that some crocodiles have salt glands in the tongue.

Despite the different locations of the salt-glands in reptiles, they are all very similar in structure. Each gland consists of densely packed branching secretory tubules radiating from a central duct (Figure 38). The central duct connects to the anterior portion of the olfactory chamber and drains to the outside via the nares, eyes or mouth, depending on the location of the gland.

The cells lining the secretory tubules are mainly columnar and have large numbers of mitochondria and lateral cell walls that are thrown into numerous interlocking folds (Figure 38). These structural features are associated with high metabolic activity and the maintenance of an osmotic gradient. These are essential for the production of secretions that are more concentrated than the blood (hyperosmotic).

Herbivorous lizards generally use their salt glands to excrete potassium, which is abundant in their food. On the other hand, marine and some terrestrial carnivorous species, including varanids, secrete mainly sodium, which is more abundant than potassium in their food and environment. At least some species are able to vary their nasal secretions according to whether they ingest high levels of sodium or potassium.

Thus goannas are able to regulate the amount of water and salts in the body by the interplay of the kidneys, cloaca and nasal glands. Urinary pellets are produced that contain little water, but large amounts of electrolytes in the form of insoluble urates. Any excess sodium that remains from sodium-linked water resorption in the kidney–cloaca system can be eliminated as concentrated nasal secretions, which again involve the loss of only small amounts of water.

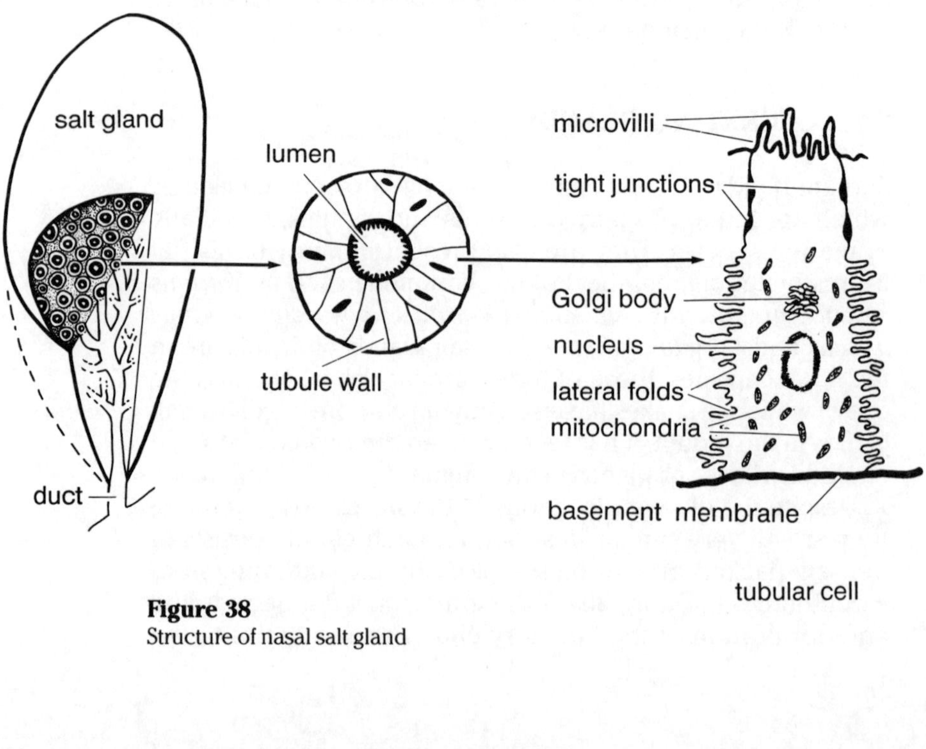

Figure 38
Structure of nasal salt gland

WATER INTAKE

Animals gain water from their environments by drinking and by taking in water present in their food. In addition, water is formed when they metabolise foodstuffs. They also exchange water vapour between the atmosphere and the lung and skin surfaces. The amounts of water that enter (influx) and leave (efflux) an animal's body can be measured by means of heavy (isotopic) water. This is water in which the hydrogen atoms are heavier than normal because of one (in deuterium) or two (in tritium) additional neutrons in the nucleus. While the deuterium atom is stable, tritium is unstable and breaks down in radio-active decay. In the laboratory, it is possible to tell the difference between these different isotopes, thus enabling researchers to use them as markers to follow changes in the quantities and the movements of water in the bodies of animals.

If isotopic water is injected into an animal it circulates and mixes with the water in the body tissues and fluids. After a few hours, the isotopic water becomes completely mixed and evenly spread throughout the body. If the precise volume of isotopic water injected is known, then its degree of dilution after mixing gives an accurate measure of the total volume of water in the animal's body. Any water that the injected animal subsequently takes in (by feeding, drinking or vapour exchange) is normal 'unlabelled' water which dilutes the 'labelled' body water. Any water that is lost by evaporation, excretion or vapour exchange is a mixture of both labelled and unlabelled water. Consequently the concentration of labelled water in the animal's body is constantly being diluted (Figure 39). Therefore, by measuring the initial dilution of the labelled water and its concentration after a known period of feeding and drinking, the amount of water that entered and left the body during that period – water turnover – can be calculated.

When an animal is in water balance (when water loss and intake are equal), the amount of body water remains constant and a simple equation accurately describes the changes in the water intake of the animal. However, more complex equations are necessary to calculate water exchange if the body water pool changes, as it does with animals that are not in water balance, or are growing.

A number of field studies have been made of the water turnover of varanids living under natural conditions. There are marked seasonal variations in water turnover in two temperate species, *Varanus rosenbergi* (Figure 40) and *Varanus varius*. The lowest daily rates of water flux are found during winter (5 millilitres of water per kilogram), and are about one quarter of the

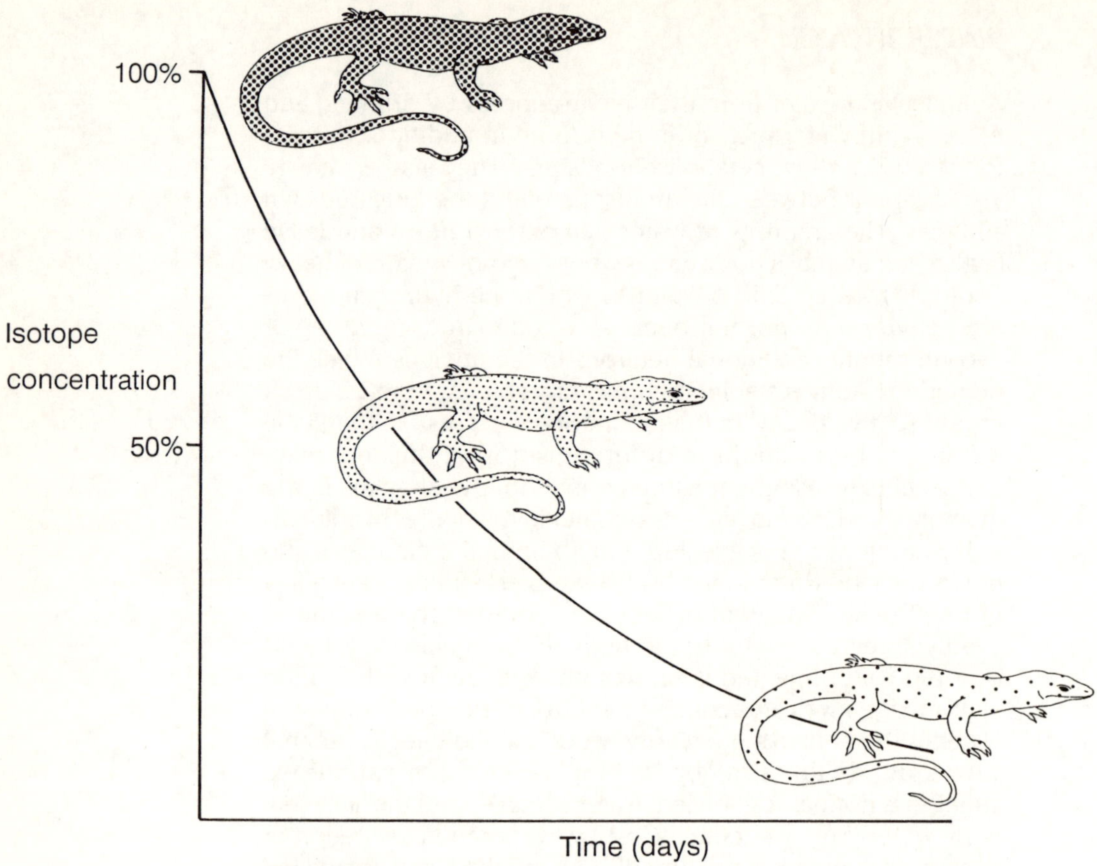

Figure 39
Measurement of water turnover by isotope dilution

summer values in both species. In winter, the goannas do not feed very often, if at all. The animals are in negative water balance and lose more water than they can obtain. About 60 percent of the water intake is derived from food and the metabolic breakdown of fat reserves and body tissues. The remaining water influx is from the exchange of water vapour between the animal and the atmosphere, both within and outside the burrow, and perhaps by some drinking.

The daily water turnover rates of both species are highest in summer, at about 22 millilitres of water per kilogram. This reflects the greater level of activity of goannas at this time, and their high rates of evaporative and excretory water losses. It also reflects their correspondingly high water intake from increased food consumption, as vapour exchange and drinking are negligible at this time.

The water turnover rates at other times of the year are intermediate between those of summer and winter, those in

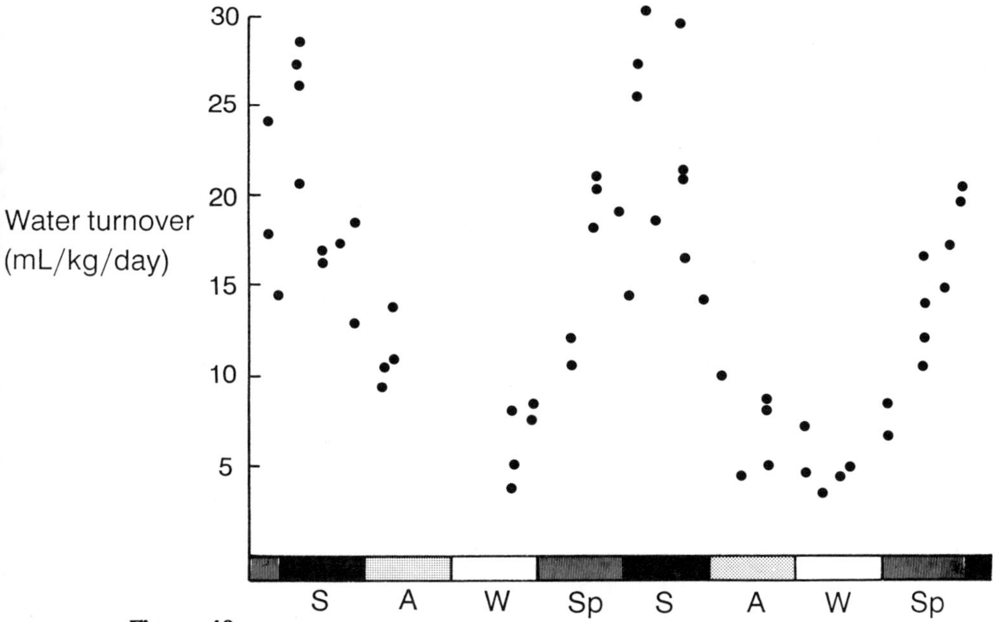

Figure 40
Seasonal rates of water turnover in *V. rosenbergi*: summer (S); autumn (A); winter (W); spring (Sp)

spring being slightly higher than those in autumn in *Varanus rosenbergi* (Figure 40). *Varanus rosenbergi* are able to maintain water balance in all seasons but winter. The seasonal variations in water turnover rates reflect changes in climate, general activity and probably prey availability. The areas over which the animals range also correlate with their water turnover rates. Goannas forage over much greater areas during summer, when they use most water, than at other times.

Other studies of field water requirements of varanids are not as extensive as those on *Varanus rosenbergi* and *Varanus varius* and are usually confined to summer or 'dry' periods. The results of these studies are shown in Table 13.

Body size has an important influence on many aspects of an animal's biology, including the surface area (Figure 33) and metabolic rate. As a consequence, it can be expected to affect water turnover. These body-size (allometric) relationships can be described by a general formula indicating the relationship between the body-size of an animal and its rate of water flux. The relationship shows that in mammals and birds, smaller animals generally have higher rates of water flux for each gram of body weight than larger animals. However, in varanids and carnivorous lizards in general, small animals have rates of water flux per gram of body weight similar to large ones (Table 13a). The ecological significance of this observation is discussed later in Energy and Food, Chapter 9.

The water turnover rates of arid and temperate varanids appear to be quite uniform, and are low compared to those of most reptiles. However, tropical species appear to have higher rates of water turnover than reptiles from other habitats (Table 13b). To some extent, this is probably due to these animals drinking, as there is an abundance of free water available in their environments. However, the higher rates of water turnover are also influenced by the higher metabolic rates of these lizards, and the much more humid atmospheres in tropical climes. Both of these factors would increase the exchange of water across the skin and lungs. More research is needed to explain the differences in the water requirements of goannas from different environments.

Table 13(a)
Water turnover rates of free-living varanid lizards in spring (sp) and summer (su)

Species	Season	Weight (g)	Percent body water	Water turnover (mL/kg/day)
V. acanthurus	sp	60	70.3	15.9
V. griseus	su	350	77.9	28.3
V. gouldii	su	493	77.0	23.5
V. rosenbergi	su	1086	76.7	22.0
	sp	1089	76.5	15.8
V. varius	su	4300	72.7	24.6
	sp	–	–	15.0
V. giganteus	su	7700	68.4	22.3
V. komodoensis	su	45000	74.9	25.5

Table 13(b)
Water turnover rates of some tropical varanids

Species	Weight (g)	Percent body water	Water turnover (mL/kg/day)
V. bengalensis	2560	76.9	60.5
V. salvator	7600	74.9	54.4

CHAPTER 9

ENERGY AND FOOD: The cost of living

We have seen that goannas are almost exclusively carnivorous and prey on a variety of animals to satisfy their energy needs. Animals require energy for all of the physiological processes associated with the normal functioning of the body. These include muscular activity, nervous transmission, growth and reproduction. This energy comes from the process of respiration, that is, the chemical breakdown (oxidation) of glucose, which produces energy, carbon dioxide and water. Glucose is derived from the basic foodstuffs, which are carbohydrates, proteins and fats.

These foods provide different amounts of energy when they are oxidised. Carbohydrates release 17 kilojoules of energy for each gram oxidised, fats yield 39 kilojoules per gram and proteins 24 kilojoules per gram. However, not all of the energy from the breakdown of proteins is available to the animal, as some is lost as nitrogen in the urine. If urea is the main waste product, only 20 kilojoules of energy can be gained by an animal from the metabolism of each gram of protein, whereas if urates or uric acid are excreted, only 18 kilojoules can be gained from each gram of protein metabolised.

Terrestrial animals all contain relatively high levels of water and protein, and reasonable amounts of fat. Thus a carnivore can obtain large amounts of water and energy from its food. In contrast, plants vary extensively in their chemical composition with regard to water, protein, carbohydrates and energy (Table 14). Grasses, in particular, can be extremely dry and fibrous during drought and yet they contain high levels of water after good rains. Because of this, carnivores are much less affected by changes in environmental and climatic conditions than are herbivores. Carnivores are what they eat. Not all of the prey material of a carnivore can be digested. Fur, feathers, some bone and invertebrate exoskeletons pass through the gut and are eliminated in the faeces. Generally between 80 and 90 percent of the total energy present in prey can be digested and metabolised by carnivores such as goannas. Herbivores feeding on fibrous diets cannot obtain as much energy from their food as carnivores.

When food materials are metabolised, they require different amounts of oxygen, and produce different amounts of carbon dioxide. Animals that are metabolising only carbohydrate produce as much carbon dioxide as the amount of oxygen used. However, if an animal is metabolising only fats, as happens during long term fasting, carbon dioxide production is only 70 percent of oxygen consumption. For protein-based metabolism, carbon dioxide production is about 80 percent of oxygen consumption. Carnivore diets provide very little carbohydrate, as the main food components are protein and some

Table 14
Composition of some dietary items.

Species	Water (% of weight)	Energy (kJ/g fresh weight)
VERTERBRATES		
Leilopisma sp. (skink)	73.6	5.16
Gehyra variegata (gecko)	74.0	5.31
Mus domesticus (house mouse)	65.9	7.57
Macropus eugenii (tammar wallaby)	69.6	6.46
INVERTERBRATES		
Insect larvae	79.1	5.02
Isopods	64.9	5.05
Beetles	65.4	8.61
Earthworms	75.1	3.53
Centipedes	71.9	6.11
Scorpions	68.9	7.54
Cockroaches	70.4	5.67
PLANT MATERIAL		
Dry grass	14.0	16.28
Lush grass	84.0	3.03
Gum leaves	57.0	9.46
Saltbush	82.0	3.75

Note: Generally 80–90 percent of the energy content of animals can be digested and assimilated by predators, however herbivores can usually obtain only approximately 50 percent of the energy contained in grass and leaves fat. This type of diet would produce carbon dioxide volumes of about 75 percent of the oxygen used.

It is quite easy to measure the amount of carbon dioxide that an animal produces in the laboratory, or the amount of energy it uses (its metabolic rate) when it is kept under artificial conditions. However, only in recent years has a method been developed to measure accurately the metabolic rate (carbon dioxide production) of free-living animals. This method is commonly called the doubly-labelled water turnover technique and is similar to the isotopic hydrogen water turnover technique described in Water Use, page 75. The technique uses water labelled with an isotope of oxygen (^{18}O) as well as water labelled with tritium (^{3}H) or deuterium (^{2}H). ^{18}O has a mass of 18 instead of the usual mass of 16 for oxygen. When ^{18}O is injected into an animal in labelled water, it is lost from the body not only in the form of water, but also as respiratory carbon dioxide produced during metabolism. By injecting an animal with both hydrogen-labelled and oxygen-labelled water it is possible to measure the rate of water loss, the amount of ^{18}O isotope lost as water and the total amount of ^{18}O lost. The difference between the total ^{18}O loss and that lost in water is the ^{18}O lost as carbon dioxide. The relative loss of isotopes is shown diagrammatically in Figure 41.

The doubly-labelled water method has been used to study the field metabolic rates and energy use of a wide range of animals, including a number of varanids. The most detailed

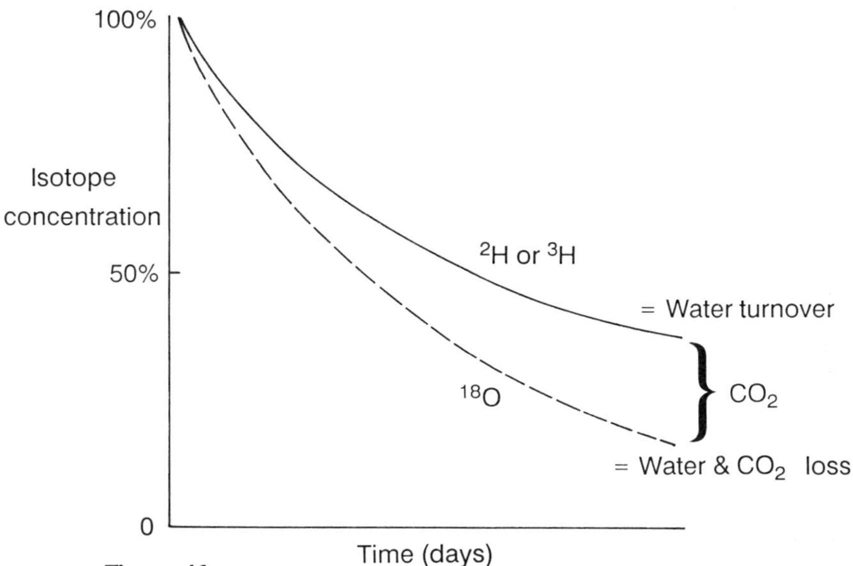

Figure 41
Measurement of carbon dioxide production using doubly-labelled water

study of energy acquisition and use by a goanna has been made on *Varanus rosenbergi*. The seasonal pattern of carbon dioxide production is similar to that of water intake. The highest metabolic rates are in summer, the lowest in winter, and they are intermediate in spring. Metabolic rates were not measured in autumn. The amount of food a goanna requires to balance its metabolic energy expenditure can be calculated if the energy content of the diet is known. In making such calculations, it can be assumed that 26 kilojoules of energy are expended for each litre of carbon dioxide produced, and that 85 percent of the energy content of prey is digested and metabolised.

These calculations show that in summer an average *Varanus rosenbergi* weighing 1 kilogram uses 112 kilojoules of energy each day and would need to consume about 22 grams of food to balance its energy expenses. In winter the same goanna would expend only 22 kilojoules each day and need only 4.5 grams of food to achieve energy balance. Even so, a goanna does not feed regularly in winter and uses more energy than it gains for much of this period. The extra energy needed in winter comes from fat stored in the body during summer. Since the water influx rates in spring and autumn are similar in *Varanus rosenbergi*, it can be assumed that the metabolic rates in spring and autumn are also similar. We can therefore assume that a 1 kilogram goanna will use 62 kilojoules per day, and consume about 12 grams of prey each day during these seasons.

A total annual energy budget for *Varanus rosenbergi* is shown in Table 15. For these calculations, it was assumed that summer and winter each last 120 days, while spring and autumn each last 60 days. Over a whole year, a 1 kilogram goanna will use 23 800 kilojoules of energy and will need to eat about 4700 grams of food to balance its energy expenditures.

Table 15
Annual energy budget of a 1 kilogram *Varanus rosenbergi*

Season	Days	Energy spent (kJ/day)	Seasonal total: Energy (kJ)	Seasonal total: Food (g)
summer	120	111.7	13400	2680
autumn	61	62.4	3800	760
winter	123	22.5	2800	550
spring	61	62.4	3800	760
annual total			23800	4750

This annual energy budget assumes that the goanna neither gains nor loses weight. If an animal is still growing, laying down fat reserves or producing eggs, it will have a higher energy expenditure and food requirement. We can calculate the total amount of energy stored in an average clutch of eggs as follows. The average clutch size of *Varanus rosenbergi* is 14 and each egg weighs an average of 25.7 grams. The average energy content of an egg is 8.1 kilojoules per gram, so the total energy content of a clutch is 2900 kilojoules. This is about 12 percent of the total energy used in a year by a non-breeding goanna. However, a lean female entering the breeding season in early spring has only 4 months in which to accumulate the energy needed to produce a clutch of eggs. Therefore, it is more realistic to compare the energy stored in the clutch with the daily energy expenditure during spring and early summer. When this is done it is found that the energy commitment in a clutch represents an increase of 32 percent over the average daily energy requirement during the breeding season. If the biochemical energy cost involved in egg production in goannas is similar to that of other lizards, then egg production represents a 40 percent increase over non-breeding energy expenditure.

To acquire the extra energy needed to produce a clutch of eggs, a female would need to capture an additional 668 grams of prey during spring and early summer. If prey or carrion are not relatively abundant during this time, it may be difficult for her to accumulate adequate energy reserves to enable her to breed successfully every year. However, if a female does not accumulate enough energy to breed in one year, she may enter winter with a greater than normal energy reserve, and this may increase her chances of breeding successfully the following year. Interestingly, some captive varanids are able to produce two clutches of eggs in a year, probably because of their abundant and regular food supply.

Goannas that acquire more energy than they need lay down reserves of body fat in a pair of fat bodies. These are located in the posterior region of the abdominal cavity (Figure 13). In lean animals, these fat bodies are small, but in some animals they can reach a substantial size. They can weigh at least 10 percent of the total body weight in *Varanus giganteus* and 7.6 percent in *Varanus rosenbergi*. Some species, including *Varanus rosenbergi* and *Varanus gouldii* also store some fat in the tail.

A comparison of the annual energy requirements of some free-living vertebrates similar in size to a Rosenberg's goanna (about 1 kilogram bodyweight), and estimates of their food and water intakes, is shown in Table 16. It is clear that the goanna is far more conservative in its use of energy and water than are 'warm-blooded' (homeothermic) mammals and birds. A goanna eats about five times its own body weight of prey each year. By comparison, a marsupial carnivore (the Eastern quoll, *Dasyurus viverrinus*) eats about 65 times its own weight, a marsupial herbivore (the brush-tailed bettong, *Bettongia penicillata*) about 70 times and a bird (the Little Penguin, *Eudyptula minor*) about 115 times its own weight annually. The goanna makes substantial savings in water and energy expenditure by allowing its body temperature to drop during the night, and by remaining relatively inactive and at a low body temperature during winter. The homeotherms pay a heavy price in energy use in order to stay 'warm-blooded'.

Most reptiles show characteristically low field metabolic rates and water turnover rates. This makes them ideally adapted to exploiting habitats lacking in energy and water, such as deserts. Ectotherms, such as varanids, are very successful in such harsh environments. Their ability to allow major variations in body temperature to occur, and their ability to remain inactive for long periods, should not be regarded as 'primitive' or 'inferior'. Instead, such strategies should be looked upon as effective alternatives to those used by homeotherms in the use of water and energy.

The majority of lizards are smaller than most mammals and birds. Eighty percent of species weigh less than 20 grams and 65 percent of species less than 10 grams. It seems that body size does not influence the water, energy and food requirements of free-living carnivorous lizards to the same extent that it does in birds and mammals. Because of this, lizards do not incur the penalties that small size imposes on homeotherms such as shrews or hummingbirds, which need great amounts of energy to remain active. Thus lizards can be very small and are able to exploit ecological niches that birds and mammals cannot use because of their high energy and food requirements.

Table 16
Comparative use of water and energy in a 1 kg reptile, mammals and a bird

	Energy (MJ)	Food (kg)	Water (Litres)
Rosenberg's goanna (*Varanus rosenbergi*) 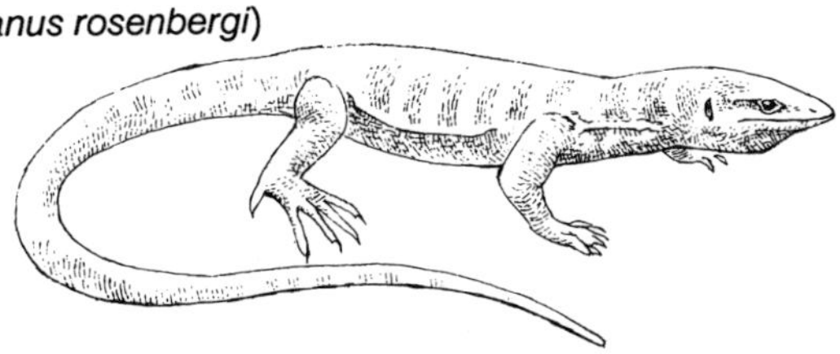	24	4.7	5
Eastern quoll (*Dasyurus viverrinus*) 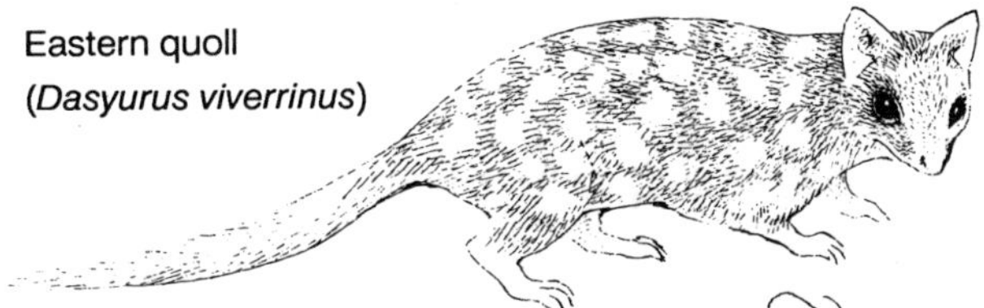	327	65.4	56
Brush-tailed bettong (*Bettongia penicillata*) 	212	68.9	42
Little penguin (*Eudyptula minor*) 	446	115	108

CHAPTER 10

CONSERVATION AND MANAGEMENT

Many reptile species are hunted and exploited for commercial purposes by humans, and varanids are no exception. Humans are also responsible for many perturbations of natural habitats such as land clearing and pollution, which also exert pressures on populations of varanids. Let us look at the conservation status of varanids on a regional basis.

AUSTRALIA

The conservation status of all Australian varanids is sound. They are protected from exploitation by state, federal and international legislation which is generally well enforced. Commercial use for skins or meat is prohibited and collection or keeping of varanids is permitted only for research purposes or by approved zoos.

Several species of goannas are used for food by Australian Aborigines, but the extent of their use and the species involved vary from region to region. They comprise only a small part of the diet of residents of coastal regions, but for desert dwellers they may make up a larger proportion of the meat in the diet than does large game such as kangaroos and emus. The survival of goannas in any area is probably not threatened by Aboriginal exploitation, as hunting for them is generally done on an opportunistic basis. Goannas are collected when encountered by the hunters who are searching for large game, or by women and children who are foraging for plants and small animals.

Goannas feature in many of the dreamtime stories of the Aborigines, are important dreamtime figures believed to occupy some of their sacred sites, are represented in many of their rock and bark paintings, and feature as tribal and personal totems. They are thus very important animals in many Aboriginal cultures.

Some goannas are killed by vehicle traffic on roads, but even the larger species of monitors (*Varanus giganteus* and *Varanus rosenbergi*) have relatively small home ranges or activity areas. Roads and heavy traffic are uncommon in most parts of Australia and most goannas do not encounter this problem.

Habitat alteration does pose a threat to some local populations. Highly developed urban areas obviously provide little or no suitable habitat for goannas. The larger species of goannas in southern Australia are generally still abundant in areas where farming has been carried out for many years, and has greatly altered the natural habitat. These species are also often seen in the outer metropolitan areas of even the larger cities.

Most species of Australian goannas have wide distributions so that habitat alterations are unlikely to occur over the entire range of the species. There are some species with restricted distributions, such as *Varanus kingorum* and *Varanus pilbarensis* but the actual distributions may be larger than are currently known. Also, both of these species exist in remote areas where the level of human activity is likely to remain low.

A different kind of human impact that poses a major threat to many species of goannas in northern Australia results from the presence of the cane toad, *Bufo marinus*. It was introduced into Queensland from South America, via Hawaii, in 1935 in a misguided attempt to control insects in sugar-cane fields. It failed to do so, and has now spread over much of northeastern Australia. It is predicted that it may reach Darwin and the vast wetlands of northern Australia before the year 2000. The toads secrete a highly toxic substance from glands located in the skin on the shoulders. This poison is fatal to large varanids and other predators which attempt to eat the toads. The toads themselves are also large animals and voracious feeders, which eat almost any type of prey and may consume the smaller species of varanids. They are believed to have already had a substantial impact on the small vertebrate fauna in large areas of Queensland, and may compete with varanids for food.

One of the few commercial uses of goannas by European settlers in Australia was the production of the patent medicine 'Goanna Oil', which supposedly was able to cure a wide range of ailments. These appear to have included almost any ailment its users wished to treat. It was believed to possess amazing properties, including the oft-repeated claim that the oil was so fine that eventually it would penetrate the glass bottle in which it was kept. In reality, the oil was a mixture of wintergreen, menthol, pine, peppermint and eucalyptus oil. Originally it did contain oil from fat bodies of *Varanus varius* but for many years has contained no oil from goannas. It is unlikely that enough goannas were ever used in its production to have had any significant effect on their numbers, even within the local area near the factory.

ASIA

The conservation status of several species of goannas in southeast Asia is strikingly different from that of Australian species. In most countries in the region there is a long tradition of exploiting monitors for food, use in traditional medicines and for skins which are used locally in the manufacture of drums and other objects. Their eggs are also highly regarded as food in many areas. This exploitation probably seriously affects only the populations of monitors in heavily settled areas.

Since skins have become commercially valuable, and highly so in the under-developed areas that comprise most of southeast Asia, huge numbers of several species of monitors have been killed to provide skins for the fashion industries of wealthy countries (*Varanus salvator, Varanus bengalensis, Varanus indicus* and *Varanus flavescens*). Little or nothing is known of the population sizes or dynamics of any of these species. The maximum sustainable rates of harvest are unknown, so it is not known whether the numbers being harvested are acceptable or not. The accuracy of the information on the actual numbers being killed is also not known. Live animals or their skins may be easily shipped across national borders, so international laws must be enforced to limit this trade.

Available figures show that *Varanus salvator* is the Asian species of the greatest commercial importance. Over 220 000 skins of this species were exported from 10 Asian countries in 1980. This number did not include over 175 000 skins of *Varanus salvator* which were exported from countries where the species does not even occur. The number of skins exported from the same 10 Asian countries during the following 5 years ranged from 425 000 to 1 300 000, and those from the countries where they do not occur or from unknown sources ranged from 52 000 to 223 000. The main source of skins was Indonesia, which exported almost 1 800 000 during the entire period. Over 750 000 were exported from Singapore, which has a total area of 600 square kilometres. Obviously the vast majority of these skins must have originated in other countries. The main countries which imported these skins were Belgium, France, Italy and Japan. There is often poor agreement between the number of skins supposedly exported and those reported as being imported.

The number of *Varanus salvator* skins which have been exported from Indonesia has increased steadily from 81 000 in 1980 to 700 000 in 1987. Most of the animals were caught by semi-professional hunters who are usually farmers or fishermen. The harvest mostly consists of animals in the age range of 3–6 years, which means they have little opportunity to breed.

Extensive hunting of *Varanus salvator* still occurs in the Philippines despite the fact that the export of wildlife products is supposedly subject to strict control. In 1985, over 80 000 skins from the island of Mindanao alone were exported to Japan. Many of these were the brightly coloured *Varanus salvator cumingi* (Plate 15). Populations of this species are decreasing in many parts of the country because of deforestation and clearing of mangrove areas. Hunting for skins, which occurs even in wildlife sanctuaries and other sparsely inhabited regions, exacerbates the situation.

The only other Asian species for which export figures are available is *Varanus indicus*. In a 10 year period, a maximum of 13 000 skins were exported (mainly from Indonesia) in any year. This species is very widely distributed in the south Pacific region, where it was apparently spread deliberately by German, Japanese and American troops in the twentieth century as a potential source of food, and as an intended pest control agent. It has, however, become a pest itself because of its predation on fowl and other beneficial animals, and is often killed for this reason. In some areas it is also used as food.

The skins of other Asian species are used by leather manufacturers but the numbers that are harvested are apparently much lower than those of *Varanus salvator*, or they are not reported. Populations of *Varanus bengalensis* in Sri Lanka were reported to be heavily reduced in the early 1950s because they were hunted for skins. This trade no longer occurs.

Similar trends have occurred in the harvest rates for skins of other large reptiles (lizards, snakes and crocodiles). However, the largest living lizard, *Varanus komodoensis,* is not hunted for its skin as the scales contain large bony inclusions (osteoderms) that make the skins unsuitable for manufacture into leather goods. In addition, *Varanus komodoensis* occurs mainly in a national park, where it is legally protected. However, most of the exploited wildlife species have had legislative controls and restrictions placed on their exploitation, though it is often impossible to enforce this legislation. Inadequate numbers of enforcement officers, lack of equipment, official disinterest or corruption, poor education levels of the local people (and often of customs officials) and the high economic value of the skins can all contribute to ineffectual legal enforcement. If, as seems likely, the high number of monitors being harvested is excessive, legal constraints seem unlikely to provide the necessary level of protection. In addition, the level of legislative protection of monitors varies greatly between the countries of southeast Asia.

The most successful means of reducing the number of monitors killed would appear to be reduction of demand, and thus the market value of the skins. This could be achieved by persuading the potential buyers of skin-based goods, such as luggage, handbags, shoes, belts and watchbands, that such purchases are ethically unacceptable. This method has proved successful in reducing the demands for other products such as furs and ivory, which when harvested in a largely uncontrolled fashion can endanger the survival of particular species. Another useful approach is the development of well designed farming programs to produce skins and thus minimise the threat to wild varanid populations. Commercial crocodile farms are already operating successfully in a number of countries. Captive monitors have also been bred successfully in recent years at a crocodile farm in India.

Another important factor which could have a major impact on monitors in southeast Asia is the massive habitat destruction occurring over much of the region. Forest-dwelling species are at particular risk from this practice, as huge areas are being cleared by logging, either legally or illegally. Additional clearing is creating more agricultural land, often in unsuitable areas. The large and rapidly increasing human populations of the region are exploiting the natural resources at an increased rate, and there is a continual demand for more land. Undeveloped areas are under increasing threat. The conservation ethic is either a totally alien concept or is of little or no concern to poor peasants when they can see opportunities to obtain more food, money or land.

The effects of pollution by industrial wastes or agricultural chemicals on monitors in this region are unknown. Far fewer restrictions on industrial pollution exist in most parts of southeast Asia than in more affluent regions of the world. Many agricultural pesticides that have been banned from use in Europe or North America are still readily available in most southeast Asian countries. Few data are available on their influence on the health or reproductive performance of monitor populations in those countries, or on other animal or human populations.

Because monitors are carnivores and are near the top of the food web, they probably concentrate and accumulate within their tissues many toxins derived from their prey.

AFRICA

The desert monitor, *Varanus griseus,* is widespread over most of North Africa, but is at very low densities in many areas due to very harsh desert conditions. Most of the area also has a very low human population, which is largely nomadic. The nomads use many parts of the monitor in their traditional medicines.

The Nile monitor, *Varanus niloticus,* and the savannah monitor, *Varanus exanthematicus,* have long been harvested throughout their distributions for their skins, for food, and for use in traditional medicines. The high commercial value of monitor skins has increased the incentive to hunt them, and more animals are being killed.

The total number of skins of *Varanus niloticus* reported to have been exported from 11 countries between 1980 and 1985 was 2 773 000, which varied from 263 000 skins in 1983 to 566 000 in 1982. Up to 210 000 skins of this species were exported in 1 year from unknown sources and from countries in which *Varanus niloticus* does not occur. Most skins were exported from Sudan, Nigeria and Mali; the main countries that imported them were Italy, France, Switzerland and the USA.

Fewer *Varanus exanthematicus* skins were exported than *Varanus niloticus* skins. The maximum number exported during any year between 1980 and 1985 was 240 000 in 1981, and the minimum was 14 000 in 1984. Most skins were exported from Nigeria and Sudan; the countries that imported most of them were Germany, Italy and the USA. Skins of both species are also used in the production of stuffed animals which are sold to tourists.

Information on population sizes and the status of all these species is very limited. The levels of legislative protection for monitors vary greatly between different countries. Despite this, there is no evident threat to any African species, as they seem to be common in protected or sparsely inhabited areas. Nevertheless, heavy exploitation of African monitors for use by the fashion industry may lead to local extinctions.

EPILOGUE

She emerged from the burrow, her cold and spent body moving stiffly and slowly onto the damp, cold earth immediately outside. She waited only briefly before moving towards a fallen tree, raising her body upon a small branch and its mat of dead leaves, escaping the cold touch of the ground. She awaited the weak warmth of the winter sun.

The clouds thickened and the wind rose. A shower of rain fell. Still she waited. A heavier shower began; she stirred herself and stolidly returned to the shelter of her burrow, seeking escape from the now pelting rain. She moved to the end of the burrow and curled up in the enlarged chamber there. The little body warmth gained from the winter sun was conserved for a while longer. Thoughts of warmer days, which would allow the insects and small lizards that made up most of her diet to become active, may have entered her brain. If so, they were wishful thoughts. The low winter temperatures would protect them for most of the months ahead as she would lie in the cool, dark burrow, waiting for days when the sun was strong enough to let her emerge and attempt to warm herself above the temperature of the winter soil. Maybe it would be warm enough to enable her to hunt and to restore her depleted body reserves. Maybe. Until then, she slept.

1. INTRODUCTION

Auffenberg, W. (1981). *The Behavioral Ecology of the Komodo Monitor.* University Presses of Florida: Gainesville, Florida.

_______ (1988). *Gray's Monitor Lizard.* University Presses of Florida: Gainesville, Florida.

Estes, R. (1984). Fish, amphibians and reptiles from the Etadunna formation, Miocene of South Australia. *Australian Zoologist* 21: 335–43.

Hecht, M. K. (1975). The morphology and relationships of the largest known terrestrial lizard, *Megalania prisca* Owen, from the Pleistocene of Australia. *Proceedings of the Royal Society of Victoria* 87: 239–49.

King, M. and King, D. (1975). Chromosomal evolution in the lizard genus *Varanus* (Reptilia). *Australian Journal of Biological Science.* 28: 89–108.

Pianka, E. R. (1986). *Ecology and Natural History of Desert Lizards: Analyses of the Ecological Niche and Community Structure.* Princeton: Princeton University Press.

2. TAXONOMY AND PHYLOGENY

Audley-Charles, M. G., Hurley, A. M. and Smith, A. G. (1981). Continental movements in the mesozoic and cenozoic. In *Wallace's Line and Plate Tectonics.* Clarendon Press, Oxford.

Baverstock, P. R., King, D., King, M., Birrell, J. & Krieg, M. (in prep). The evolution of the Varanidae: Microcomplement Fixation analysis of serum albumins.

Becker, H. O., Böhme, W. & Perry, S. F. (1989). Die Lungenmorphologie der Warane (Reptilia: Varanidae) und ihre systematisch-stammesgeschichtliche Bedeutung. *Bonn Zoologische Beitrage* 40: 27–56.

Böhme, W. (1988). Zur Genitalmorphologie der Sauria: funktionelle und stammesgeschichtliche Aspekte. *Bonner Zoologische Monographien,* Nr. 27: 1–176.

Branch, W. R. (1982). Hemipeneal morphology of platynotan lizards. *Journal of Herpetology* 16: 16–38.

Carter, D. B. (1990). Courtship and mating in wild *Varanus varius* (Varanidae: Australia). *Memoirs of the Queensland Museum* 29: 333–8.

Estes, R. (1983). Sauriaterrestrie, Amphisbaenia. In Wellnhofer, P. (ed.) *Encyclopaedia of Paleoherpetology*. Stuttgart, Verlag.

_______(1984). Fish, amphibians and reptiles from the Etadunna formation, Miocene of South Australia. *Australian Zoologist* 21: 335–43.

Fejevary, G. J. (1918). Contributions to a monography on fossil Varanidae and Megalanidae. *Annals Museum National Hungarici* 16: 341–467.

Gillespie, J. M., Marshall, R. C. and Woods, E. F. (1982). A comparison of lizard claw keratin proteins with those of avian beak and claw. *Journal of Molecular Evolution* 18: 121–9.

Holmes, R. S., King, M. and King, D. (1975). Phenetic relationships among varanid lizards based upon comparative electrophoretic data and karyotypic analyses. *Biochemical Systematics and Ecology* 3: 257–62.

King, D., King, M. and Baverstock, P. (1991). A new phylogeny of the Varanidae. *Mertensiella* 2: 211–19.

King, M. (1990). Chromosomal and immunogenetic data; a new perspective on the origin of Australia's reptiles. In *Cytogenetics of amphibians and reptiles: Advances in life sciences*. Birkhauser Verlag. Basel.

_______ and King, D. (1975). Chromosomal evolution in the lizard genus *Varanus* (Reptilia). *Australian Journal of Biological Science*. 28: 89–108.

Mertens, R. (1942). Die Familie der Warane (Varanidae). Erster Teil: Allgemeines. *Abhandlungen der Senckenbergischen Naturforschenden Gesselenschaft* 465: 1–116.

Pregill, G. K., Gauthier, J. A. and Greene, H. W. (1986). The evolution of helodermatid squamates, with description of a new taxa and an overview of Varanoidea. *Transactions of the San Diego Natural History Museum* 21: 167–202.

3. FEEDING

Auffenberg, W. (1981). *The Behavioral Ecology of the Komodo Monitor*. University Presses of Florida: Gainesville, Florida.

_______ (1988). *Gray's Monitor Lizard*. University Presses of Florida: Gainesville, Florida.

Bellairs, A. d'A. (1949). Observations on the snout of *Varanus*, and a comparison with that of other lizards and snakes. *Journal of Anatomy* 83: 116–46.

Cissé, M. (1972). L'alimentation de Varanidés au Senegal. *Bulletin de l'Institut Français d'Afrique Noire*, Serie A 34: 503–15.

James, C. D., Losos, J. B. and King, D. R. (1992). Reproductive cycles and diets of goannas (Reptilia: Varanidae) from

Australia. *Journal of Herpetology* 26: 128–36.

King, D. and Green, B. (1979). Notes on diet and reproduction of the sand goanna, *Varanus gouldii rosenbergi*. Copeia 1979: 64–70.

_______, Masini, L. and Robinson, R. (in prep). Diet and reproduction of *Varanus scalaris* and *V. tristis*.

Losos, J. B. and Greene, H. W. (1988). Ecological and evolutionary implications of diet in monitor lizards. *Biological Journal of the Linnean Society*. 35: 379–407.

Pianka, E. R. (1986). *Ecology and Natural History of Desert Lizards: Analyses of the ecological niche and community structure*. Princeton: Princeton University Press.

Rieppel, O. (1979). A functional interpretation of the varanid dentition (Reptilia, Lacertilia, Varanidae). *Gegenbaurs Morphologie Jahrbuk*, Leipzig 125: 797–817.

Shine, R. (1986). Food habits, habitats and reproductive biology of four sympatric species of varanid lizards in tropical Australia. *Herpetologica* 42: 346–60.

Smith, K. K. (1986). Morphology and function of the tongue and hyoid apparatus in Varanus (Varanidae, Lacertilia). *Journal of Morphology* 187: 261–87.

Weavers, B. W. (1989). Diet of the Lace Monitor Lizard (*Varanus varius*) in south-eastern Australia. *Australian Zoologist* 25: 83–5.

4. BREEDING

Auffenberg, W. (1981). *The Behavioral Ecology of the Komodo Monitor*. University Presses of Florida: Gainesville, Florida.

Bredl, J. and Horn, H-G. (1987). Über die Nachzucht des australischen Riesewarans *Varanus giganteus* (Gray, 1845). *Salamandra* 23: 90–6.

Cowles, R. B. (1930). The life history of *Varanus niloticus* (Linnaeus) as observed in Natal, South Africa. *Journal of Entomology and Zoology* 22: 1–31.

Carter, D. B. (1990). Courtship and mating in wild *Varanus varius* (Varanidae: Australia). *Memoirs of the Queensland Museum* 29: 333–8.

Darevsky, I. S. and Auffenberg, W. (1973). Growth records of wild recaptured Komodo monitors (*Varanus komodoensis*). *HISS News-Journal* 1: 41–2.

Horn, H.-G. and Visser, G. (1989). Review of reproduction of monitor lizards (*Varanus* spp.) in captivity. *International Zoo Yearbook*. 29: 140–50.

_______ and _______ (1991). Basic data on the biology of

monitors. *Mertensiella* 2: 176–87.
King, D. and Green, B. (1979). Notes on diet and reproduction of the sand goanna, *Varanus gouldii rosenbergi*. *Copeia* 1979: 64–70.
_______, _______ and Butler, H. (1989). The activity pattern, temperature regulation and diet of *Varanus giganteus* on Barrow Island, Western Australia. *Australian Wildlife Research* 16: 41–7.
King, D., Masini, L. and Robinson, R. (in prep.). Diet and reproduction of *Varanus scalaris* and *V. tristis*.
_______ and Rhodes, L. (1982). Sex ratio and breeding season of *Varanus acanthurus*. *Copeia* 1982: 784–7.
Riley, J., Stimson, A. F. and Winch, J. M. (1985). A review of squamata ovipositing in ant and termite nests. *Herpetological Review* 16: 38–43.
Rookmaaker, L. C. (1975). The history of some Komodo dragons (*Varanus komodoensis*) captured on Rintja in 1927. *Zoologische Mededelingen* 49: 65–71.
Shine, R. (1986). Food habits, habitats and reproductive biology of four sympatric species of varanid lizards in tropical Australia. *Herpetologica* 42: 346–60.

5. GENERAL BEHAVIOUR

Cowles, R. B. (1930). The life history of *Varanus niloticus* (Linnaeus) as observed in Natal, South Africa. *Journal of Entomology and Zoology* 22: 1–31.
Green, B. (1972). Water losses of the sand goanna (*Varanus gouldii*) in its natural environments. *Ecology* 53: 452–7.
_______ and King, D. (1978). Home range and activity patterns of the sand goanna, *Varanus gouldii* (Reptilia: Varanidae). *Australian Wildlife Research* 5: 417–24.
Horn, H.-G. (1985). Beitrage zum Verhalten von Waranen: die Ritualkampfe von *Varanus komodoensis* Ouwens, 1912 und *V. semiremex* Peters, 1869 sowie die Imponierphasen der Ritualkampfe von *V. timorensis timorensis* (Gray, 1931) und *V. t. similis* Mertens, 1958. *Salamandra* 21: 169–79.
Murphy, J. B. and Mitchell, L. A. (1974). Ritualized combat behavior of the pygmy mulga monitor lizard, *Varanus gilleni* (Sauria: Varanidae). *Herpetologica* 30: 90–7.
Pianka, E. R. (1986). *Ecology and Natural History of Desert Lizards: Analyses of the ecological niche and community structure*. Princeton: Princeton University Press.
Shine, R. (1986). Food habits, habitats and reproductive biology of four sympatric species of varanid lizards in

tropical Australia. *Herpetologica* 42: 346–60.

Stanner, M. and Mendelssohn, M. (1987). Sex ratio, population density and home range of the desert monitor (*Varanus griseus*) in the southern coastal plain of Israel. *Amphibia-Reptilia* 8: 153–64.

Vernet, R., Lemire, M., Grenot, C. J. and Francaz, J. M. (1988). Ecophysiological comparisons between two large Saharan lizards, *Uromastix acanthinurus* (Agamidae) and *Varanus griseus* (*Varanidae*). *Journal of Arid Environments* 14: 187–200.

_______, _______ and _______ (1988). Field studies on activity and water balance of a desert monitor *Varanus griseus* (Reptilia, Varanidae). *Journal of Arid Environments* 15: 81–90 .

6. THERMAL BIOLOGY

Bowker, R. G. (1984). Precision of thermoregulation of some African lizards. *Physiological Zoology* 57: 401–12.

Cowles, R. B. and Bogert, C. M. (1944). A preliminary study of the thermal requirements of desert reptiles. *Bulletin of the American Museum of Natural History* 83: 261–96.

Green, B., King, D., Braysher, M. and Saim, A. (1991). Thermoregulation, water turnover and energetics of free-living Komodo dragons, *Varanus komodoensis*. *Comparative Biochemistry and Physiology* 99A: 97–101.

Huey, R. B. (1974). Behavioural thermoregulation in lizards: importance of associated costs. *Science* 184: 1001–3.

Hutchinson, V. H. and Larimer, J. L. (1960). Reflectivity of the integuments of some lizards from different habitats. *Ecology* 41: 199–209.

King, D. (1980). The thermal biology of free-living sand goannas (*Varanus gouldii*) in southern Australia. *Copeia* 1980: 755–67.

_______, Green, B. and Butler, H. (1989). The activity pattern, temperature regulation and diet of *Varanus giganteus* on Barrow Island, Western Australia. *Australian Wildlife Research* 16: 41–7.

Murrish, D. E. and Schmidt-Nielsen, K. (1972). Exhaled air temperature and water conservation in lizards. *Respiration Physiology* 10: 151–8.

Sokholov, V. E., Sukhov, V. P. and Chernyshov, Y. M. (1975). A radio telemetric study of diurnal fluctuations of body temperature in the desert monitor (*Varanus griseus*). *Zoological Zhurnal* 54: 1347–56.

Vernet, R., Lemire, M., Grenot, C. J. and Francaz, J. M. (1988). Ecophysiological comparisons between two large Saharan lizards, *Uromastix acanthinurus* (Agamidae) and *Varanus griseus* (Reptilia, Varanidae). *Journal of Arid Environments* 14: 187–200.

Weavers, B. W. (1983). Thermal ecology of *Varanus varius* (Shaw), the Lace Monitor. Unpublished Ph.D. thesis, Australian National University: Canberra.

Wikramanayake, E. and Green, B. (1989). Thermoregulatory influences on the ecology of two sympatric varanids in Sri Lanka. *Biotropica* 21: 74–9.

7. RESPIRATION

Becker, H. O., Böhme, W. and Perry, S. F. (1989). Die Lungenmorphologie der Warane (Reptilia: Varanidae) und ihre systematisch-stammesorschichtliche Bedeteung. Bonn *Zoologische Beitrage* 40: 27–56.

Bellairs, A. d'A. (1949) Observations on the snout of *Varanus*, and a comparison with that of other lizards and snakes. *Journal of Anatomy* 83: 116–46.

Bennett, A. F. (1972). The effect of activity on oxygen consumption, oxygen debt, and heart rate in the lizards *Varanus gouldii* and *Sauromalus hispidus*. *Journal of Comparative Physiology* 79: 259–80.

_______ (1973) Blood physiology and oxygen transport during activity in two lizards, *Varanus gouldii* and *Sauromalus hispidus*. *Comparative Biochemistry and Physiology* 46A: 673–90.

Glass, M. L., Wood, S. C., Hoyt, R. W. and Johansen, K. (1979). Chemical control of breathing in the lizard, *Varanus exanthematicus*. *Comparative Biochemistry and Physiology* 62A: 999–1003.

Heisler, N., Neumann, P. and Maloiy, G. M. O. (1983). The mechanism of intracardiac shunting in the lizard *Varanus exanthematicus*. *Journal of Experimental Biology* 105: 15–31.

Perry, S. F. (1983). Reptilian lungs. Advances in anatomy, embryology and cell biology. Berlin, Heidelberg: Springer Verlag.

Webb, G., Heatwole, H. and DeBavay, J. (1971). Comparative cardiac anatomy of the reptilia 1. The chambers and septa of the varanid ventricle. *Journal of Morphology* 134: 335–50.

Wood, S. C. and Johansen, K. (1974). Respiratory adaptations to diving in the Nile Monitor Lizard, *Varanus niloticus*. *Journal of Comparative Physiology* 89: 145–58.

8. WATER USE

Bradshaw, S. D. and Rice, G. E. (1981). The effects of pituitary and adrenal hormones on renal and post-renal reabsorption of water and electrolytes in the lizard *Varanus gouldii* (Gray). *General and Comparative Endocrinology* 44: 82–93.

Braysher, M. and Green, B. (1970). Absorption of water and electrolytes from the cloaca of an Australian lizard, *Varanus gouldii* (Gray). *Comparative Biochemistry and Physiology* 35: 607–14.

Dryden, G., Green, B., King, D. and Losos, J. (1990). Water and energy turnover in a small monitor lizard, *Varanus acanthurus*. *Australian Wildlife Research* 17: 641–6.

Green, B. (1972). Aspects of renal function in the lizard *Varanus gouldii*. *Comparative Biochemistry and Physiology* 43A: 747–56.

_______ (1972) Water losses of the sand goanna (*Varanus gouldii*) in its natural environments. *Ecology* 53: 452–7.

_______, King, D. and Butler, W. H. (1986). Water, sodium and energy metabolism in free-living Perenties *Varanus giganteus*. *Australian Wildlife Research* 34: 589–95.

_______, _______, Braysher, M. and Saim, A. (1991). Thermoregulation, water turnover and energetics of free-living Komodo dragons, *Varanus komodoensis*. *Comparative Biochemistry and Physiology* 99A: 97–101.

Lifson, N. and McClintock, R. (1966). Theory of use of the turnover rates of body water for measuring energy and material balance. *Journal of Theoretical Biology* 12: 46–74.

Rice, G. E., Bradshaw, S. D. and Prendergast, F. J. (1982). The effects of bilateral adrenalectomy on renal function in the lizard *Varanus gouldii* (Gray). *General and Comparative Endocrinology* 47: 182–9.

Vernet, R., Lemire, M. and Grenot, C. (1988). Field studies on activity and water balance of a desert monitor *Varanus griseus* (Reptilia, Varanidae). *Journal of Arid Environments* 15: 81–90.

9. ENERGY AND FOOD REQUIREMENTS

Dryden, G., Green, B., King, D. and Losos, J. (1990). Water and energy turnover in a small monitor lizard, *Varanus acanthurus*. *Australian Wildlife Research*. 17: 641–6.

_______, _______, Wikramanayake, E. and Dryden, K. (1992). Energy and water turnover in two tropical varanid lizards, *Varanus bengalensis* and *V. salvator*. *Copeia* 1992, 102–7.

Gales, R. and Green, B. (1990). The annual energetics of Little Penguins (*Eudyptula minor*). *Ecology* 71, 2297–2312.

Green, B., Dryden, G. and Dryden, K. (1991). Field energetics of a large carnivorous lizard, *Varanus rosenbergi*. *Oecologia* 88: 547–551.

_______, King, D., and Butler, W. H. (1986). Water, sodium and energy metabolism in free-living Perenties, *Varanus giganteus*. *Australian Wildlife Research* 34: 589–95.

_______, _______, Braysher, M. and Saim, A. (1991). Thermoregulation, water turnover and energetics of free-living Komodo dragons, *Varanus komodoensis*. *Comparative Biochemistry and Physiology* 99A: 97–101.

Lifson, N. and McClintock, R. (1966). Theory of use of the turnover rates of body water for measuring energy and material balance. *Journal of Theoretical Biology* 12: 46–74.

Nagy, K. A. (1982). Energy requirements of free-living iguanid lizards. In *Iguanas of the World: Their behavior, ecology and conservation*. Noyes Publications: Park Ridge, New Jersey.

Vernet, R., Lemire, M., Grenot, C. J. and Francaz, J. M. (1988). Ecophysiological comparisons between two large Saharan lizards, *Uromastix acanthinurus* (Agamidae) and *Varanus griseus* (Varanidae). *Journal of Arid Environments* 14: 187–200.

10. CONSERVATION AND MANAGEMENT

Auffenberg, W. (1981). *The Behavioral ecology of the Komodo Monitor*. University Presses of Florida: Gainesville, Florida.

Erdelen W. (1991). Conservation and population ecology of monitor lizards: the water monitor *Varanus salvator* (Laurenti, 1768) in south Sumatra. *Mertensiella* 2: 120–35.

Gould, R. A. (1980). *Living Archaeology*. Cambridge University Press: Cambridge.

Luxmoore, R., Groombridge, B. and Broad, S. (eds). (1988). Significant trade in wildlife: a review of selected species. In *CITES Appendix II. Reptiles and invertebrates*. International Union for the Conservation of Nature and Natural Resources: Gland, Switzerland.

Vernet, R. and Chamaillard, M. (1984). Les varans. In *Especes menacees et exploitees dans le monde; guide pratique pour leur connaissance et leur identification*, ed. F. de Beaufort, Secretariat de la Faune et la Flore: Paris.